ARCTURUS ASTROLOGY

A Consciousness Accelerator for
The Aquarian Age

Ashtara

Published by Tara Rising in 2021
www.tararisingpress.com

Prepublication Data available from the National Library of Australia.

ISBN: 978-0-9803178-3-1 (pbk)
Also available as an ebook
ebook ISBN: 978-0-9803178-4-8

Cover design: In2Art Designs

Book design and publishing assistance by
Publicious P/L
www.publicious.com.au

Also written by Ashtara
Gaia, Our Precious Planet
Tara, Emissary of Light
The Great Cosmic Joke
A Treasure Trove of Gems
A Crack in the Cosmic Window
Your Recipe for Empowerment through Spiritual Astrology
Volumes One, Two and Three
Esoteric Astrology, The Astrology of the Soul
I Am an Experiment, An Extraordinary Spiritual Adventure
The Magdalen Codes, Reclaiming Ancient Wisdom

DEDICATION

To my students and readers without whom this book could not have been written. Your support, love, and belief in me and my work carried me through many a dark night. Thank you from the bottom of my heart.

CONTENTS

INTRODUCTION

My ability to access higher-frequency thought began through health necessity. Through regular daily meditation, I learned to still and calm my over-active and nervous brain. Teaching myself to write in a daily journal with my left, non-dominant hand assisted this developmental phase. Through relaxing, sensing full body and mind feelings, and attentive listening, I opened my heart and mind to frequencies of thought normally filtered out of awareness. I developed a loving, and trusting relationship with my many interdimensional trainers, both celestial and of the stars, who taught me so much.

During this early training period, I learned to experience astrology from an intuitive and feeling perspective, awakening many soul memories of ancient knowledge as I did so. Because I am a sceptic, I always sought proof. I didn't believe in astrology until I applied it to myself.

Then I knew, without a shadow of a doubt, that it worked as a psychological and self-healing tool when practiced and applied to self. I healed myself from my ailments and through my faith in astrology and in Divine Spirit guiding me. I learned many invaluable techniques to accelerate my journey and that of my students into higher consciousness, through self-awareness.

My healing tools have been a combination of evolutionary and esoteric astrology, soul psychology and spirituality. New languages needed to be learned. Not only an astrological language but also one of esoteric psychology, metaphysics and spirituality, and once learned, uniting all three into simple coherent expression.

In my experience, astrology is a universal language of the soul, and as such can be used to uncover, and to become aware of, the subconscious psychology that drives us into creating and living life as we do. When we learn to observe ourselves from a 'watcher' perspective realising we are the creators of our reality, we can commit to doing whatever is necessary to bring about effective and positive internal change. In so doing, we learn to harmonise our subconscious motivations and allow our intuitive higher mind to guide our lives. This process enables union between our internal feminine and internal masculine's objective and analytical mind.

Following an initial astrology reading in 1987, my oldest daughter, Christine, gave me my first astrology book. I did not have a birth chart at that time however I didn't need one to read and understand the book. *Spiritual Astrology, Your Personal Path to Self-fulfillment* by Jan Spiller and Karen McCoy.

When reading it, revelation followed revelation. Truth was felt and known. The written words in the section on the pre-natal eclipses described my inner struggle to make sense of my reality. With a pre-natal lunar-eclipse in Gemini specific words resonate as truth, or otherwise, within me. My body and soul recognise and respond to key words.

In the author's text they make mention of Babylon, a time referred to in the Bible when various tribes were split into different languages, no longer able to communicate with each other. At the early age of seven I studied the Bible, reading it from cover to cover unable to comprehend the words but feeling an intuitive connection to the teaching, much of it symbolic.

The chapter in The Book of Daniel interested me. According to my Bible, a revised standard King James version, King Nebuchadnezzar, King of Babylon, besieged Jerusalem. He commanded his chief eunuch to bring some of the people of Israel, both of the royal family, and of the nobility, youths without blemish, handsome and skilful in all wisdom, endowed with knowledge, understanding and learning, and competent to serve in the kings' palace, and to teach them the letters and language of the Chaldeans.

The king assigned them a daily portion of rich food which the King ate, and of the wine he drank. After three years they were required to stand before the King. Among these were Daniel and Hananiah. They were given new names. Daniel, he called Belteshazzar and Hananiah he called Shadrach. There were two others in this group.

To these four youths, God gave them learning and skill in all letters and wisdom. Three of them, Shadrack, Meshack and

Abednego had developed their spiritual abilities to the extent that, when they were thrown into King Nebuchadnezzar's fiery furnace, they were able to consciously raise their vibration sufficiently so they weren't burned by the heat force. Their faith and trust in God saved their lives.

Possibly their learning and skills developed because of knowing of a language code that connects the biblical alphabets of Hebrew and Arabic to modern chemistry. Scientist and author, Gregg Bradon, refers to this code as *The God Code* in his book of the same name. What it means in simple terms is that the letters of God's ancient name are encoded as the genetic information in every cell of life.

Or perhaps they knew that all life is connected to God consciousness, and this loving energy resides in every cell of our bodies. This is my understanding. Shadrach was my first spiritual guide. He made himself known to me during a meditation early in my astrological studies and helped me to learn and integrate the material.

I persisted with my learning knowing it was the tool I needed to save my life from heading further down a dark abyss. I learned that accessing higher-frequency life-force awareness is an organic process, a natural ability every child has at birth. This ability usually atrophies when language is learned, possibly because of the cultural belief that if something cannot be put into words it isn't real.

Hence, the universal language of telepathy, and of feeling and sensing knowledge and wisdom arising within us from subconscious memories, is discarded for something more tangible.

Through integrating the wisdom shared by the authors of the *Spiritual Astrology* book, I learned to understand a common thread running through all things in the universe. I combined the knowledge with my super-sensitivity, now a blessing rather than the hinderance it had been. Thank you, Jan Spiller and Karen McCoy, for your combined words of wisdom. They established a foundational path that still fulfils, sustains and nourishes me.

Through personal interaction with the planetary energies via a particular meditation given to a large group of students at an astrological conference in the mid-1990s I learned to invoke the planets. This meditation has been a most useful tool in understanding and healing my discordant psychological energies. I am forever grateful to the brilliant Swiss Astrologer/Psychologist Verena Bachmann who led the conference group through the meditation.

Years later, I created a simple version of *Invoking the Planets*, sharing it with many of my astrology groups. It has been the most amazing tool for my evolutionary and astrological growth, and for that of my students. It embraces the feminine intuitive and feeling perspective.

Verena also took the large group through a second meditation that I only used once. Why? Because in that meditation was revealed my past, present and future in clear, symbolic detail. I intuitively knew not to repeat the experience because it felt true and complete. What I experienced during that meditation manifested. I refer to it as *My Inner Home* meditation. It too has aided my students on their evolutionary spiritual growth path.

As a result of these experiences, I learned to know the planetary energies from a different perspective from that of the logical, analytical left brain. During the meditations I learned to telepathically communicate with, and to befriend, the archetypal energies of the planets. They became my friends and guides. I understand them to be spirit beings of high evolvement.

To receive their clear communication, I needed to completely still my mind so I could be fully present and alert in each now moment. I tuned into their frequencies, and the suggestions they gave me were immeasurably helpful. I felt their energies as supportive and loving. Through them I learned that there is no star, planet, luminary, asteroid, or human that does not have a spirit life.

Throughout the three decades of teaching astrology as a tool for developing higher consciousness through self-awareness, I continued working with my associated interests – metaphysics, spirituality and esoteric psychology. These combined studies, and applied practices, provided me with initiatory tools, techniques and exercises I'd learned and taught in previous lives.

Through immersing myself in the above study, and the application to my life of all I was learning, I began to understand the great secret of astrology: that every planet is a living consciousness and we have the ability to connect with the consciousness of each of the planets and stars and to integrate their positive characteristics and qualities.

I learned to comprehend their inner connections within me, and, because of my sensitivity, I could feel their movements in my brain. I also learned to communicate telepathically with their archetypal energies and have taught my students to do the same. In my mind, the planets have a spiritual job to do, and that is to activate the raising of human consciousness, according to their individual vibratory patterning. They do this through transits to our natal chart.

According to Carl Jung, astrology is the psychology of the Soul. I agree. Our birth chart reveals subconscious psychological issues we have incarnated to consciously realise and alchemise because they no longer serve our evolutionary growth. Chief among these are fears, judgements, revenge, separation from love, lack of self-trust, social convention and tradition.

When you give them energy, i.e., serve them, they can become extremely destructive. I learned their origin usually stems from previous incarnations and perpetuate through situations experienced in childhood. Their lower vibrations act as a magnetic attractor.

These planetary interactions led to being trained by celestial and evolved star light beings who taught me to view life and my world through a new lens, the lens of higher consciousness. Because I had learned to trust, the training was exciting, explorative, and hugely fulfilling.

Throughout this developmental phase, I became a scribe for celestial and highly evolved star beings. As I was sitting at my computer preparing to send Part One and Two of this book to a friend who offered to edit it, a message came through me.

We, a collective consciousness residing in finer vibratory fields to those of planet earth, contacted Ashtara many decades ago, asking if she would be our scribe. She challenged us, and when she felt we met her discerning principles, she agreed.

I, The Tibetan, asked if she would write a book for me. She agreed, on the proviso she couldn't do it in the same intellectual way as Alice A. Bailey, who was my scribe in the earlier years of the 20th century. I agreed. The following is the final book of a trilogy. It has taken enormous courage for Ashtara to write this book, and she resisted doing so for many years. Neither she nor I have any need for recognition. In devotion to the Great Plan for human evolution of consciousness, Ashtara completed a mammoth task. We thank her.

"Who is The Tibetan?" you may ask. He is Ascended Master Djwhal Khul. "How do I receive these messages?" you may also ask. That story follows.

As a consequence of devoted spiritual and astrological practice, dedicated to raising not only my own consciousness but also that of my students, I was trained by the Arcturians, enlightened star beings, to learn a new astrology system. This system takes those who have worked through most of their birth-chart incarnational psychological challenges to a higher level of consciousness. I learned that nothing can stop those who hearts burn with genuine passion for the truth.

Through deep love for planet earth and her continual evolution, I realised that Eternal Spirit is a current in our lives that can guide when we develop the awareness of it doing so. I learned that our bodies are a biological circuitry system, designed to project our heart-felt vision into our material world when we allow the process.

Learning to realise, and to accept, that we are a cell in the body of the One Prime Creator whose consciousness is now awakening in the human family, is part of our purpose for being human at this time in our evolutionary history, the Age of Aquarius.

Through the experiential learning process, I feel and intuit deeper understanding and meaning via my right brain. I then engage my left brain to make sense of it. When 'connecting the dots' in my mind and body, I feel an energetic 'wash' of comprehension and integration.

I also feel the chakras of my etheric energy body resonating to the revealed truth. Excitement comes next, and often I telepathically hear the words of my Sirian healer/guide, Athenia, saying "Precisely". I rejoice, because I've cracked another code in a giant jigsaw puzzle of life experiences. Another jig-saw puzzle piece can now be inserted into its rightful place.

This information contained in this trilogy was given to me via the way I learn and integrate, through experience.

According to esoteric astrology and my present understanding, it is the work of the zodiacal influences to evoke the emergence of the *spiritual will* of all souls and personalities. The energies of the planets, and the signs of the zodiac, produce an effect upon unevolved human as well as the advanced human, galvanising the chakra centres above the diaphragm into activity. This process enables individuals to respond to the radiation and activity of the Spiritual Hierarchy.

The goal of evolution for humanity is to become consciously and livingly aware of the nature of these energies and begin to know, and to use them.

The first thing to do is to become consciously self-aware, through focused and applied study of the zodiacal and planetary influences, and begin to use their energies for carrying out your Soul's purpose.

These influences I have described simply and clearly in my trilogy of Spiritual Astrology work books titled *Your Recipe for Empowerment through Spiritual Astrology.* My purpose for writing that trilogy was to provide practical tools to aid the spiritual evolution of each student who studies them in conjunction with their personal natal (birth) chart. This trilogy of easy to read workbooks is available through my website.

The planetary transits to placements in the birth chart are the influential energies that activate the soul into desire for spiritual evolution. The personality may ignore the inner prompts and learn the hard way, through illness, accidents, life traumas such as marriage break-ups, financial issues etc.

All are 'wake-up' calls to pay attention to, and become aware of, one's inner world and psychological motivations and projections. The birth chart clearly shows the individual's life lessons to be learned and the gifts and talents available to be used to overcome the psychological blockages that cause weakness and disempowerment.

In effect, my teachings have provided my students with the tools to develop emotional intelligence, body awareness, psychological self-mastery and spiritual growth.

In this book I will do my best to provide the stepping stones, the tools to aid your personal enlightenment and transmutation of form into spirit.

I serve as an emissary for the Council of Nine from Sirius, and a scribe for the Arcturians and The Record Keepers, of which Djwhal Khul (D. K) is the spokesperson. I gladly serve you as a catalyst, paradigm shifter, guide and friend. I will guide you to the realisation that you are the path – the Light and the Way. I ask you to open the door of your Christ heart to know your true self.

Arcturus Astrology will accelerate your journey to the Light. Many changes to your perceptions will take place. Sometimes you may need to breathe, open your heart and mind, and relax.

When you realise that your personality is a one-time composite of soul fragments, gathered together by your soul group, or over-Soul, prior to incarnation, you will begin to open your mind to further possibilities. Your over-Soul, on a lighter and higher dimension to that of the third dimensional earth, can see further than your personality.

Your DNA exists not only on the physical genetic level, but also on multi-dimensional soul levels. To access these multi-dimensional levels requires focused inner spiritual growth work.

Your birth chart is a coded map that provides access to past life gifts and talents, current life challenges, strengths and weaknesses including health issues, and offers glimpses of our potentiality and your future, should you earn the right to move up the evolutionary ladder to higher consciousness.

As you read the pages of this book, I invite you to have the courage to be an open-minded child-like explorer. Keep questioning as you read, accepting there are no "right" answers. We all have choices. Choose to be fully self-aware in each present-moment time. What messages is your body giving you as you read? How does your heart feel as you read? These feelings are your truth. Choose to feel your truth.

Though I have been led and inspired by many authors, the major thrust of this work has been received from the archetypal levels of consciousness as light enhanced pictographic images within my mind, and clear telepathic communication from Light Beings, both celestial and of the stars. This is not information over which I have no control, but rather a mutually co-operative exchange engineered on my behalf as a result of the direct, conscious application of esoteric Law and understanding to my life.

I offer myself in service to readers through the medium of the written word. Some of my views may be new to you. As I learned through experience, I ask you do the same. Apply the teachings and if they work for you, continue with them. Use your intuition, that comes from heart-felt feelings, to be your guide.

I do not purport to share absolute truth, only truth that has worked for me. Is there such a thing as absolute truth? I think not! This book is about my journey with spirit, and written for people of spirit. If these words support even just a few readers to become more consciously self-aware, they will have served their purpose.

May your heart be opened, and your mind expanded to allow emergence of your soul's feeling truth and intuitive knowing. May you know and be guided by the Eternal Presence within.

Ashtara
Tamborine Mountain, Queensland,
Australia
July, 2021

PART ONE

Spider Woman Weaves a Web

To several of the First Nation people of the American Southwest, Grandmother Spider Woman is the feminine energy of the creative force. From the Galactic Centre, she spins a magical web of a peaceful world. Her intricately beautiful designs catch the morning dew, sparkling in the sunlight. Some get caught in her sticky, non-linear web unable to see beyond it. Continually creating, she demonstrates the expansiveness of the Eternal Plan.

Throughout many years of regular meditations, I was taken on extraordinary space adventures to experience a number of different stars. I was asked by celestial guides to record and write of these experiences. Many of them are written about in the two earlier books of this trilogy, *I Am an Experiment, An Extraordinary Spiritual Adventure* and *The Magdalen Codes, Reclaiming Ancient Wisdom*. The process of learning *Arcturus Astrology* arose from being taught this system in a board room on Arcturus where there were other students present.

As soon as I came out of a meditation, and while still in the frequency of it, I wrote of my experiences with my left, non-dominant, hand to ensure my analytical brain didn't take over. I've experienced wonderful adventures into unknown multi-dimensional regions. These regions felt familiar, as if I'd experienced them many times before.

This book provides the answer to my persistent question:

What is the cosmic system that activates further human evolutionary spiritual growth when we have worked through the psychological issues depicted in our birth charts?

Remembrance of teaching in ancient mystery schools in Peru, Bolivia, Egypt and Greece, of being an astrologer in Tibet, and other lives as an Australian aborigine and a north American Indian, have enabled this book to be written. These memories have been pivotal in raising my consciousness. Some of these wondrous experiences are written about in the two previous books of this trilogy. I've also remembered lives once lived in India, and in Lemuria.

When living in Atlantis, I realised a calamity was imminent and warned of it. I was not heard. With heartache and pain, I observed that land mass and its civilisation collapse, and made a vow: to do whatever I could to prevent such a calamity from occurring again.

I know, without a shadow of doubt, that the Way of the Heart, the Way of rightful use of energy, is the only way to prevent another such global calamity. A calamity on planet earth will affect the solar system and beyond.

Taught telepathically this life in high alchemy teachings (internal energy practices) by Spiritual Masters and groups of advanced star beings from Sirius, Orion, Antares, the Pleiades and Arcturus, I've practiced and shared them with students and clients.

Planet Earth is going through a transition phase, a birthing process. Years ago, I acted as mid-wife to the birthing of the etheric light body of Gaia. It was an amazing experience, and I intuitively knew what to do to assist the process. Now approaches her birthing time on the physical level. What happens during a physical birth? The pregnant female feels the urge to clean her space. Her waters break, and contractions begin.

*

As students advance to become spiritual initiates, they learn to recognise, through bodily feelings and senses, the influential cosmic energies. These can be experienced as a vibration that registers in one or other of the seven energy centres (chakras) as a revelation of a particular type of light conveying a specific colour, a particular note, and/

or a directional sound. An initiate on the path to self-realisation can experience all three.

The weakening psychological issues involving traumas and lower vibrational energies contained within the four lower chakras, those below the diaphragm, need to be consciously identified, accepted, responsibility taken for creating them, and worked through before the Christ Light and Magdalen Codes can emerge from within the individual's consciousness. The four chakras above the diaphragm – the heart, throat, third eye and crown, are basically receiving centres. These four receiving centres activate the three lower centres – base of spine, sacral and solar plexus/spleen.

Throughout the initiates journey to the light of higher consciousness, experiences similar to a crucifixion and resurrection take place. Not as a physical crucifixion, as per Jesus' experience into his light body through resurrection, but one whereby Soul crucifies a certain amount of dark, dense psychological dross gathered over many lifetimes of experience, followed by a psychological rebirth into a new and higher awareness, enabling the integration of greater light. This evolutionary growth process often occurs under a Pluto transit.

When these evolutionary processes are addressed through self-awareness and deep inner psychological work, and the positive energising results are attained as a constant in an individual's life, the centres above the diaphragm become radio-active, dynamic and magnetic. An inner experience of a reversal of Soul's journey around the zodiacal constellations takes place. I experienced this to be the case.

In Ascended Master Djwhal Khul's words:

As more people realise this truth, astrology will be seen in a new light. The purpose for incarnation into the 3D plane of experience is to grow and evolve spiritually and psychologically, through self-awareness. This process results in the creation of finer light frequencies permeating the human Soul. Both the human spirit and the Soul facilitate this growth.

Karma is accrued through the universal Law of Attraction, often referred to as the Law of Cause and Effect. Every thought and deed have an effect, either positive or negative. The direction of Soul growth is clearly evident in an individual's birth chart and requires knowledge, intuition and experience to reveal truth. Some astrologers take a more academic approach, focusing on the intricacies of the technology. Intellect does not reveal truth because it denies feeling and intuition, the holistic approach.

Our scribe has had a great deal of experience navigating other dimensions of reality. We encouraged her to write of them to aid comprehension. From her perspective, she had astrology karma to work through. She was able to use her feminine nature to delve deeply into cause and correct the karma, learning much along the way.

*

We are moving at an accelerated pace from 3D consciousness to the prophesied 5D consciousness necessary for the manifestation of the Golden Age of light and peace, a Camelot. Many people are unaware of this process. Increasing numbers are consciously making the connection to internal Spirit and willingly initiate the necessary psychological changes. Millions of spiritual seekers are now walking this path. The wave of Cosmic Light is growing daily as more people step on board the consciousness train.

Among the many levels of assistance available to us are some advanced extra-terrestrial cultures that have been monitoring and guiding humanity throughout aeons of time. Some of these cultures belong to the Galactic Federation of Light. They value all life, and work towards furthering the Divine Plan, the Great Work for human evolutionary growth. They develop, organise and guide projects using telepathy and dream communication.

One of the primary guidance groups is the Arcturians, a collective of star beings from Arcturus, a bright star in the Bootes constellation. According to physicist and author, J.J. Hurtak, in his extraordinary book *The Book of Knowledge, The Keys of Enoch,* Arcturus is our local mid-way Light station and our closest centre for soul mapping. Arcturus, or "Ash" means "The Good Shepherd".

The Arcturians were the group who telepathically communicated my first book in 1996, *Gaia, Our Precious Planet.* They reconnected in 2007 as their second response to my persistent question:

What cosmic bodies influence the next developmental level of human consciousness evolution when our zodiacal and solar system no longer challenge?

The Arcturians first response to my above repeated question came at the turn of the millennia.

Early October 2000, during an afternoon meditation I clearly experienced my personality ascending into a pillar of cosmic white Light while at the same time witnessing a galactic aspect of my oversoul descend from the Light to occupy the available space.

Immediately following the experience, I had the presence of mind to write the time of the "walk-in". It was 2.23 pm, the time of Barbara's birth. Ashtara "walked-in" to a healthy sixty-three- year old body. This experience is described in detail my *I am an Experiment* book. I now had another birth chart to work with and through. Was this a Cosmic Joke? I took it seriously, advancing my inner Soul work from a different perspective.

The Arcturians second response came seven years later. During 2007, I was taken up to Arcturus by Archangels Gabriel and Uriel, where I was taught, and fully experienced with all my senses activated, the cosmic system I have named *Arcturus Astrology.*

The Arcturian's teachings can be likened to shamanism, the ancient body of energetic knowledge. It has served as an intermediary between the realms of the physical and the spiritual, leading people to self-discovery, self-realisation and spiritual love.

The Acturians are the observers of, and givers to human kind. They make themselves known to receptive humans to aid their consciousness evolution, and our collective evolution into greater Light. This evolutionary process into higher consciousness is being accelerated in all fields of human endeavour.

The world as we know it is rapidly changing through evolutionary necessity. No longer can we desecrate planet earth with poisonous substances, psychological negativity and manufactured garbage. No longer can greed, financial control and power by a few manipulate the many. We each have a choice to change and evolve, or resist and suffer.

Following the publication of *The Magdalen Codes* book in 2018, I was guided to relocate from the city to the country. Towards the end of the house clearing process, I found my original 2007 Arcturus transmissions and notes relative to that book, and to this, its sequel.

In the Arcturus transmissions, some of them written about in *The Magdalen Codes* book, there are repeated reference to the *Rainbow Warriors.* These *Rainbow Warriors* refer to individuals who have done a great deal of inner work to purify their personal psychology and who now live life from a higher, broader and wiser perspective. They have awakened their soul's memory of truth, and realise their purpose for incarnation.

Living from their hearts, they have learned to bring into balance the right and left hemispheres of their brain. With an open heart, they listen to, and feel into their inner guidance, and act on it. They have demonstrated great courage because it takes courage to face the truth of one's greater, and darker, side. They have chosen love over fear. *Live life in love* is their guiding light. It can also be yours. Many children born since 2000 embody this rainbow consciousness. They are our future.

The Rainbow Warriors, and the Rainbow children, so named because of the higher level of consciousness contained within the seven love-infused rainbow colour rays of cosmic light they embody and emanate, are creating a new world on Earth, a new Camelot. Their committed service is to share their love and their light to all within their field of influence. Their journey has not been easy, and is ongoing. Yet, the joy and gratitude they feel for taking it is worth the pain experienced. They pave the way for others to follow.

Should you feel a resonance to my words, I invite you to commit to accelerating your personal evolutionary journey into greater light and divine love. Divine love is quite different from human person-to-person love. It is the realisation, understanding and integration that, within each and every cell of your body and mind is the Eternal Presence of Prime Creator. The inner journey of opening up to, and re-discovering this Presence requires a loving and open heart, self-awareness, mindfulness and the right use of will. Encouraging the heart to govern your mind is a necessary step on the journey.

The purpose for writing this book is to fulfil Djwhal Kuhl's request of me to write a book for him, and also to do whatever I can to raise your consciousness to a level that enables the creation of a global Camelot, a Golden Age of Peace. Should my words stir a cord of remembrance, and a desire in your Soul to join me, I welcome you with open arms and a warm and loving heart. The joy that comes from taking this journey far exceeds any pain that accompanies it.

Welcome aboard the consciousness train. It will take you into many unknown territories and challenge you to be the best, most loving, noble and open person you can be.

I have been given the privilege, responsibility and good fortune to steward the Arcturus Astrology system into public awareness. The way I do so is to provide appropriate and nourishing soul food to those who want to learn, through the written and spoken word.

I accept the commission with humility and gratitude. I will do the best I can. I share my story in the hope your Soul will remember Its heritage. Throughout this book are many transmissions received from the Arcturians and D.K as spokesperson for The Record Keepers.

The creation and building of a new global Age of Peace - a Camelot, will take time. It will be accomplished. Of this, I have no doubt.

My profound love for Gaia, our precious planet earth, and humanity upon her, drives me forward.

*

Consciousness - An Astrological, Psychological and Spiritual Perspective

We are surrounded by a cosmic system of energy. This subtle field of energy is in constant vibration and the rate of its speed manifests matter. Everything in our universe is made of this subtle energy emanating from Source, the place of the One Great Mind. The substance of this energy is Divine Love.

This infinite energy is stepped-down via specific stars, constellations and planets, each acting as power stations. Programmed to receive and transmit particular energy packages that contain different qualities, all living things on Earth receive these stepped-down transmissions. Each human is a storehouse of a specific transduced energy package.

Our Soul fully cognizant with the system and prior to incarnation, chooses our personality based on evolutionary purpose. Spiritual and esoteric astrology is a study that identifies this operational system, providing tools to harmonise our opposing internal psychological forces.

This psychological and spiritual evolutionary development becomes more effective when we become aware of the subtle energy currents guiding our lives. These cosmic vibrations are experienced through the body's senses and organs of perception. The high frequencies of love, joy and gratitude are felt in the heart chakra. Sensitive people feel their own, and other's shadow psychological energy, and weaken because of it unless self-aware enough to deflect it.

Suppression and denial of our own shadow psychology leads to illness, disease and premature death. Self-awareness of psychological density and the specific thoughts that create it, and the willingness to change old patterns of thought and behaviour into uplifting and energising ones, require constant mental vigilance from an observer perspective. Every psychological expression creates a different vibratory tone within the human body, perceived as a frequency.

The Light of higher consciousness can be likened to a switched-on light globe. Insert a new light globe into the correct fitting, plug the appliance into an electrical socket, turn on the switch, and the light globe comes on. There is light where there once was none. You can see more clearly.

Consciousness with self-awareness, is the same. All connections must work before we can switch on our Power Source. We must be connected by feeling to our heart and Prime Creator's love within it.

Consciousness can be likened to electricity. When plugged into a power source, electricity runs through a cord activating a device. Whether the device is a computer, light bulb, or vacuum cleaner, electricity is needed to fuel the device.

In the human body, when plugged into Source energy, divine love and the light of higher consciousness runs through the synapses of our brain, meridians, chakras and blood vessels, activating the body and mind into self-realisation. When we perceive a disconnection from Source we can act as robots, willingly allowing others, such as governments, corporations, religious and educational institutions, and the media, to think and act as their agendas dictate.

We were created as unique and wondrous light beings, filled with Prime Creator's energy essence - full consciousness and infinite love. We are a bundle of vibrating energy. We can choose to identify and use our God/Goddess given energy to its greatest capacity by living our lives in joy, radiance and loving kindness, or not.

We each carry seeds of Prime Creator's love in every cell in our bodies. It is the divine essence of pure love. When these seeds are nurtured, they grow into full flowering.

All living things are evolving. Our multi-universes are governed by Prime Creator's loving vibration. Source love-based consciousness is the energy of life. This intelligent love energy is transduced through the millions of solar systems via an infinite series of Suns. A Sun is a star that radiates its own light. Planets receive the light of the Sun.

Planet Earth and we humans are in the evolutionary process of becoming a star, a Sun in our own right. Our personal evolutionary journey is to become the greatest light we can be, through embracing and emanating the frequency of divine love. When in-love with the Creative Force within, and following its heart-felt direction, our lives transform.

When, through self-awareness, we harmonise our internal polarities our external world reflects our inner change. Our Solar System planets, and the zodiacal signs provide a clear pathway to spiritual evolution and ascension into greater light.

When consciously working with astrology from this perspective, we accelerate our journey to the Light. When we harmonise our psychology through dedicated devotion to our spiritual evolutionary path, our experience of planetary energies changes. We no longer unconsciously re-act to them. Instead, we consciously use their energy to advance our spiritual and psychological growth. Knowing there is always more growth and learning to experience, my passionate and repetitive question asked internally and constantly of my Soul and spiritual guides, was:

What cosmic system is available that continues to promote human consciousness evolution after we have worked through our Solar System's planetary challenges and themes depicted in our birth charts?

My meditation experiences into finer vibratory realms to experience and define the new Cosmic System I've named *Arcturus Astrology* was the answer. It has taken years to de-code my experiences and to make sense of them. I offer my findings with no attachment as to how you receive them.

I recorded my galactic training experiences immediately I came out of them, in as much detail as I could. I share them in the hope they will help you make sense of your reality, and bring you into a higher state of consciousness through many 'light bulb' moments of self-awareness.

Accelerated evolution into higher frequencies of Cosmic Light is taking place. Science is changing its perspective on human evolution and finally science and ancient mystical and esoteric wisdom teachings are meeting. The same knowledge is expressed using different language. The understanding is the same. Our Universe is infinite. Created by a Supreme Intelligence, the essence of this Intelligence is within every living creature, known as consciousness.

*

As mentioned earlier, language divides human races. Symbolism unites because it is universal. Astrology is a symbolic language. Each symbol tells a story. By learning to feel into the symbolic story codes, truth emerges. The emphasis is on feeling.

Our soul has memories, created through heightened emotional experiences. Memories contain past experiences, often connected to other incarnations. They can become activated during planetary and cosmic cycles. Our souls know the cosmic system. When it's the cosmic time for soul memories to emerge, our body responds to the new vibration.

Our earth and solar system have entered a different area of space, experiencing different vibratory patterns. The frequencies of the Age of Aquarius are arousing soul memories connected to the Age of Leo, a time when Atlantis collapsed, approximately 12,000 years ago. The memories of this catastrophe are hidden in our souls' memory. Many humans are accessing these memories. Hence the explosion of technology. And also, the explosion of higher consciousness. Ancient knowledge and connecting psychological patterns are being activated.

The influence of the Aquarian Age energy - of Cosmic Light - will ensure spiritual evolution of our human species. How we each experience our future depends entirely on present-moment thoughts and actions.

By focusing our mind only on shadows, we could create another Atlantis. Negativity produces lower vibrations that permeate all they touch. Thought creates the reality we individual experience. Collective thought creates collective reality.

Higher vibratory energy, filled with the light of higher consciousness, self-awareness and divine love, is the pathway to spiritual ascension and evolution. The practice and application of astrology to one's psychological life is a tool to accelerate the journey. This is how I have used, and taught my students to use, astrology.

When operating life through higher consciousness we positively affect the lives of all living things within our field of influence. This process is accelerating. Our world will never be the same because we humans cannot ever be the same.

Astrology is a pathway to ascension and Christ/ Buddha/God consciousness. For two thousand years it was kept secret. The Church of Rome dictated it be abolished probably because it provided the means whereby individuals could gain personal sovereignty and psychological self-mastery that led to God consciousness. The Church created a hierarchical system whereby God/ Goddess realisation had to be experienced through an intermediary of the Church, i.e., a priest. Astrological knowledge is now available to all seekers.

Our brain is hard-wired, like a computer. Every thought causes a movement in the brain. Cosmic forces connect to the brain and all forces are vibratory. Cosmic forces travel in the form of waves, faster than light. The brain stores information containing old psychological software based on parental, societal, religious and family conditioning.

We live our lives according to this old software until such time as our soul decides to "wake up" to truth, our soul's' truth. This "wake-up" call usually comes through hardship, distress, illness and emotional upheaval. It seems to be the only way we humans awaken. It need not be so. We can live life an easier way by using astrology.

Levels of consciousness can be divided into three main sectors: Subconscious, (a repository of unconscious soul memories), conscious, and super-conscious (a state of being that arises through awakening, becoming conscious of and healing, ancient soul memories).

Spiritual Masters such as Jesus and Buddha became super-conscious humans. Our collective human race is moving from an unconscious and non-self-aware state of being into a conscious one. Cosmic forces are accelerating a major growth phase. We can pay attention or suffer. We each have free will choice. This choice is to develop inner light – the light of cosmic consciousness, or to stay in the dark, ignorant of our divine nature.

Astrology has been around since the birth of humans. Just imagine being without all technological devices. Electricity hadn't been discovered. Fire hadn't been discovered. The wheel hadn't been discovered. What did humans do each night? They spent time investigating the stars.

From thousands of years of observation and correlation they devised a system of knowledge. This knowledge evolved through each Astrological Age. An Astrological Age is 2,592 years long. We are now evolving through the Age of Aquarius. This is the Age of new technology, new ideas and huge advances in human consciousness evolution.

Each ancient culture formed its own understanding of astrology, devising different models to do so. The basic understanding is the same: cosmic forces are continually bombarding all living things, and these cosmic forces influence our state of being. We humans have the intelligence to learn to co-operate with these natural forces.

Not all the energy of the Universe vibrates at the same speed or frequency, or the same level of consciousness. Energy that is totally aware of itself moves faster, so manifestation is quicker. As we humans develop greater self-awareness, we seek to develop even higher levels of consciousness. Life is speeding up into faster evolution. There is an evolutionary revolution taking place, a non-violent one. Our time of amnesia is over.

We are all born with a number of different qualities and characteristics, developed during previous lives. In childhood we learn attitudes, habits and beliefs based on our parental and societal conditioning. One of our most important tasks in life is to uncover and develop as many of our positive qualities as possible so we can realise the full range of our potential. The planets and the twelve zodiacal signs, and the archetypal energies they symbolise, support and guide our self-discovery journey.

Because Earth is situated in the heavens between Venus (feminine principle - magnetic) and Mars (masculine principle - electric) we are electro-magnetic human beings and attract people and situations into our lives in order to learn about ourselves. Each human being is unique. No one else thinks like you do.

Your birth chart is a map of your unique consciousness and Soul's purpose for incarnation. Each of the glyphs and lines have meaning and provide a means for objectively viewing the intangible dimensions of self. You are not a victim of your birth chart.

You have the free will choice to cooperate with your own divine intelligence and, with determination, insight, emotional intelligence and mental management, move through weaknesses as they arise. How? By becoming conscious of them. When the light of conscious self-realisation sparks the emotional body, the emotional charge is lessened and eventually fades into insignificance.

Key Words

The science of energy, of consciousness, was given to the people on earth in early Atlantean times. Cosmic energy simply is, and available at all times to use. One simply needs to tune into the appropriate consciousness wave.

Scientists with an interest in ancient wisdom teachings will soon, or may have already, discovered that we humans have an energetic system through which zodiacal, planetary and cosmic waves move. Astrology will then be put to the test. Currently, it is only by feeling, sensing and experiencing the energy that we can know for certain how the system works. Electrical static in the brain is created when shadow psychology abounds. Key words trigger the body into remembrance of truth.

Thoughts and emotions are energy in motion. Vibrating at different frequencies according to their content, they create our reality. We can change our reality through self-awareness.

I've learned to view a birth chart as a symbolic representation of an individuals' unique energetic circuitry system. Each placement in the birth chart can be likened to an acupuncture point, each point representing a psychological theme.

An acupuncturist places a needle into a blocked energy point in the human body temporarily freeing up a psychological blockage, often connected to an elemental imbalance. The elements used in astrology are fire, earth, air and water. There is a link between these elements to ancient Alchemy, and their translation to modern chemistry. In alchemy, the element of Air connects to Nitrogen, Water to Oxygen, and Fire to Hydrogen.

The Arcturians have the following to say:

The Fire of Consciousness

In the beginning was the word – and the word was of God. The word was love infused thought. The pure vibration of this heart-felt thought created sound waves within the great ocean of etheric substance. From this vibratory pattern, particles of particulate matter combined to form the creation of the world. This is how creation occurs. Within this seeming nothingness a seed is planted. The seed requires the infusion of Spirit to make it grow. The original seed was an idea in the mind of Prime Creator. Why was the idea formulated? Where did it come from?

In the beginning there was stillness and seeming nothingness. Imagine a fine mist with nothing moving within that mist. The mist is. The mist is impregnated with the radiance of a sunbeam – fire. The fire is the spark of consciousness. The mist is water. Fire (consciousness) can be likened to a bolt of lightning. It disturbs the calm and stills mist. Nothing can ever be the same again.

Where did the bolt of lightning, the fire of consciousness, come from? Consciousness is the fire of Spirit. It is omnipresent and permeates all of creation. It causes creation. It requires a still mind before it can do its work effectively. The ripples become waves. The consciousness beam can be likened to a pebble, thrown into a still pond. Ripples occur. When the fiery impulse contains great potency, the ripples become waves and there is greater disturbance. Is this disturbance felt, or observed?

Imagine the mist as the human mind. The fire of spirit impregnates the mist of the mind and disturbance results. The fiery consciousness spark, coming from Source, changes the structure surrounding the pond. What if there was no structure?

The ocean of mist is everywhere and is in everything. It is seemingly nothingness. Does the nothingness perceive itself?

It requires the spark of consciousness to do so. The consciousness perceives. Does the mist feel the result of the impregnation? If the mist is the human mind, then it must. The mist becomes aware of itself.

*

Part of our human evolutionary growth involves harmonising our elemental nature. It's wise to bear in mind that our breath can serve as a healing tool to activate ancient soul memories.

During the mid-1990s I had many weekly healing sessions with a breath worker, who helped me access, through regular, deep circular breathing, past-life psychological blockages evident in my chart that prevented present-time creative expression. The added intake of oxygen to the brain enabled the revelation of these memories. Through awareness of the issues revealed, I was able to work through, and release them. I used the outer planetary transits to aid my journey.

When a transiting planet activates a birth chart placement, the new energy flows into the existing energy points according to sign, house and aspect configurations. Activation of the circuitry system occurs. Buried, unconscious issues and their themes arise seeking conscious realisation and integration. This energetic process can be felt by the experiencing individual.

Sounds carry a frequency. Words, letters, numbers, thoughts and emotions have a frequency. When the birth-chart placement is understood, the individual can use their astro/psychological knowledge and self-awareness to use the transiting cosmic energy infusion as does an acupuncturist.

Prior to applying to the birth chart 'ignition' point (the exact aspect – whether conjunction, square, opposition, or other), the self-aware individual can realise an opportunity is available for greater self-realisation. He or she can take appropriate measures to release any blocked energy.

When unconscious of subconscious motivations and thoughts, the incoming energy wave activates this lower frequency. Agitation occurs and sparks fly. A symbolic fuse is blown. The unconscious individual reacts once again to a past dysfunctional psychological pattern. The emotional reaction created the electrical fuse. Cosmic energy flow becomes blocked. Lack of psychological self-awareness created the problem.

The many planetary and cosmic energies are in constant motion, beaming energy and forming specific geometric patterns within the circuitry system of the receiver. The individual's thoughts and emotions have an impact on the circuitry. The natal chart can be viewed as a depiction of the individual's circuitry system, and when read as such, becomes a scientific tool for research.

The soul has a purpose. That is to bring the one-time personality construct into a state of unity, rather than of separation. Dysfunctional use of psychological energy created unconsciously through past and present life experiences creates separation between body, mind and spirit.

Conscious Soul infusion can occur through application of astro-psychology to one's life. It is often through emotionally charged situations and the committed desire to never, ever, repeat shadow psychology because of the self-created consequences, that the desire to embark upon the self-healing journey into higher consciousness begins.

I accepted the soul mission to work through many subconscious psychological issues in one life, rather than taking three of more lifetimes as seems to have been the normal. It made sense to me to accelerate my spiritual journey into greater light so I could be of greater service to others, supporting them to identify and work through their issues.

To develop a deeper understanding of astrology, it helps to know that the planets and stars are living, evolving entities, as are we humans. Each human, planet, star, constellation and the vastness of space itself are conscious evolving beings.

Every being receives, stores and transmits energies, the frequencies of which depends upon the amount of light carried and transmitted. Where the light is great, the frequency is fine and the radiance bright.

Each being can be likened to a cell within the great 'body' of Prime Creator. This energy permeates all of creation. Source energy is infinite and forever evolving. As electro-magnetic humans, we exist in an energy ocean of God/Goddess consciousness. When we fully realise this, life changes.

Each solar system has a central hub, a brilliant Sun containing huge energetic force and radiation. Planet Earth exists in a third dimensional energy field. Other solar systems operate in higher dimensional realms. Other light beings exist in different dimensions to that of our 3D world.

When we raise our consciousness sufficiently, we can communicate with them. This evolutionary developmental process happened to me. I have become a scribe for higher dimensional light beings who transmit valuable information through me. I write and share the information.

*

The following transmission, and those that follow in this chapter are from The Elohim, Creator Gods. They refer to the Seven Rays of Creation.

The Ancient Mystery Schools

The ancient Mystery Schools were conducted in many different parts of planet Earth during Lemurian and Atlantean times. During these times, initiates and disciples on the path to spiritual ascension participated in varying and diverse training procedures.

They were chosen to attend these schools of higher learning and were only admitted when they had demonstrated their willingness to move beyond their comfort zone of thought, emotionally re-active behaviour and dysfunctional attitudes.

Great Masters from higher dimensional realms would come to assist the training of these initiates and disciples. They would make their presence felt during relaxed meditation. And, so it is today.

These civilizations died out because of the great imbalance of human dysfunctionality. Negative thought forms abounded, and the material life became more important to the masses than their spiritual life. Civilizations destroy themselves, or are destroyed, when this imbalance occurs.

This almost happened to Planet Earth recently. However, the scales have now turned, the pendulum is swinging in the opposite direction and balance will occur. It may take some time. In the meantime, the knowledge taught and practised by the ancient civilizations mentioned above needs to be brought forward into the hearts and minds of people on Earth. Our channel is a vehicle for this.

Unbeknown to her, she has been teaching many of the principles taught in the ancient mystery schools. She is being awakened now so that these teachings reach more and more people.

The information will be presented in a simple way so that it is easily understood intuitively. We are presenting the female intuitive point of view. This information was taught at the sacred Mystery Schools of Machu Picchu, high in the Andean Mountains.

At that place, the energies are so refined that access by the Master teachers was always easy. This was the main female Mystery School on the planet at the time, and much of the knowledge learned there is ready to be made known to those with ears to hear, eyes to see and an open heart to feel.

Should you respond to the following information, and if it rings true for you, then experiment with it, play with it, feel into it, work with it and then accept it as your truth. For others not drawn to the information there are many other avenues for you to explore.

In the ancient days, the Science of Energy was considered to be one of the most important of sciences. Astrology was the subject used to understand, define and qualify specific energies emanating from various sections of our Universe. Esoteric Astrology is the study of these energies and how they impact upon all living things.

Man/woman, being the fourth kingdom of nature, has the ability to develop an understanding of these energies by learning how to feel their subtle flow within the body. The body is both the receiver and the transmitter of these energies. It is a finely tuned instrument designated specifically to work in harmony with the cosmic forces that permeate it on a daily basis.

Astrology is the tool that allows individuals to consciously access and understand this subtle energy system. When one feels and experiences something one knows it to be true, otherwise it is simply book knowledge. Our channel has been teaching people how to do this for many years.

Once the energies of the Solar System are understood and mastered, the individual is able to use the energy consciously and constructively. It is then that finer and subtler energies from outside your Solar System become apparent. At this stage, the initiate is ready to move on in spiritual growth. Evolution to a higher understanding takes place. It is from this point that we begin this major work.

The Seven Colour Rays of Creation

Following is a brief introduction:

The Rays are the energetic emanations, or expressions of Divinity and are the builders of all that exists in space. They, as living, evolving Light Beings, are concerned with the evolution of consciousness within not only the human kingdom, but all kingdoms of nature. They work in developing human consciousness via the solar system, the zodiacal constellations and other cosmic sources.

Esoteric Astrology and Psychology are based on an understanding of these Rays that are potent radiating streams of energy forming from their beginning in an atom. Every living entity is endowed with consciousness.

For a period of approximately 26,500 years humans devolved from higher states of consciousness into amnesia, unaware of their personal psychology, and the fact that, resident within their bodies is a soul and spirit.

Since our solar system moved into the higher frequencies of The Aquarian Age, a time period of approximately 2,500 years duration, and a new millennium, humans are in the process of waking up to their subconscious motivations and seeking to evolve again into cosmic, or Christ/God consciousness – our true nature. This process requires a great deal of inner work.

The following lengthy transmission explains more about the Rays. They form a vital part of the Arcturus Astrology system.

In the beginning there was The One. This One, the Source of All That Is, existed in timeless space. The One consists of energy without form. At some stage in the creation story The One chose to experience itself and to do this needed to create another aspect of self.

An energetic transference took place and there were two. The two, as independent entities, merged and experienced a union from which a third was created. We now have the principle of the triangle. Symbolically, this is the Father, Mother and Child.

Energy was stepped down from the One, the Source. Separate streams of energy were formed. These streams of energy were in a sense separate yet still connected to The One. The Source is in the other two energy streams and was their Creator however the two felt separate yet part of the greater whole.

From this triangle of energy more streams of energy were born. These formed into the Seven Rays. These Rays are living entities of tremendous power and force, although not in a form as you know it.

These Seven Rays permeate your Solar System through the constellations of the zodiac. They permeate other Solar Systems as well. Yours is not the only Solar System. There are many more. Your central star, the Sun, is one of the many that orbit around Alcyone, the central Sun of the Pleiadian system within your constellation of Taurus.

Your system of astrology was given to you through the Mystery Schools in early Atlantean times. In your world today, many of the ancient teachings have been lost however the old knowledge will now be brought forth. It is time. The student of astrology who chooses to move beyond the known and is willing to enter into a deeper intuitive understanding, will find the forthcoming information challenging. This is as it is.

In early Atlantean times, both the intuitive and the rational, logic sides of the brain were recognized as having great validity and value. It was only in later times of that period that the logical, rational left-brain constructed thought patterns gained prominence. Intellect became more important than intuition. It was this factor that resulted in the decay of the civilization.

Of course, this process took time, aeons. That state of being is currently upon your planet now. You value logic and intellect above intuitive, conceptual symbolic knowledge. This is what has led to your decay.

The light at the end of the tunnel is that there are more and more people in positions of influence who now trust their intuitive process and who are courageous enough to allow their inner knowledge and wisdom expression.

The Mystery Schools, called so because of the seemingly mysterious information that was studied and taught, were highly regarded, and it was a great honour to be chosen to be a student. This situation will again become prevalent on planet Earth. It may take a little time.

Once the initiate was admitted into the Mystery Schools, much training had to take place to ascertain the level of commitment and the strength of the will of the initiate. A rigorous training period of three months was undertaken as a probationary period. If the individual failed to fulfil the curriculum they were asked to leave and to re-apply at a later time, should they choose to do so. There was no judgment, only acceptance of where the individual was at in his/her evolutionary development of consciousness.

Those reading this story are now ready to embark on that period of training, including the author. She will say, "Oh – Oh! Another Cosmic Joke"! We are the Elohim and we enjoy creating scenarios where we can have fun. Learning is fun and the more fun we have as we learn the easier it is to integrate the teachings.

The first part of the initiates' training was to develop an understanding of how the Seven Rays function through the zodiacal signs, and, for this part of the exercise, we are indebted to the Tibetan Master Djwhal Khul (D.K.) who already has this information in circulation. Please understand that the Seven Rays are seven entities of a very high vibration and they have such tremendous power that to encounter them direct would 'blow your circuitry system'. They step themselves down through the signs thus:

Ray One: *The Ray of Will or Power, through Aries, Leo and Capricorn.*
Ray Two: *The Ray of Love and Wisdom through Gemini, Virgo, and Pisces.*
Ray Three: *The Ray of Active Intelligence through Cancer, Libra and Capricorn.*
Ray Four: *The Ray of Harmony through Conflict through Taurus, Scorpio and Sagittarius.*

Ray Five: *The Ray of Concrete Science through Leo, Sagittarius and Aquarius.*
Ray Six: *The Ray of Idealism and Devotion through Virgo, Sagittarius and Pisces.*
Ray Seven: *The Ray of Ceremonial Order through Aries, Cancer, and Capricorn.*

The challenge for you now is to work with this system, to feel the energies and to develop an understanding of why this is so.

The next part of the exercise is to develop an understanding of the nature of the Seven Rays, those illumined beings that are part of The One, as are you. The difference is in the frequency. As you develop an understanding of the energy of your Solar System and beyond, and apply the understandings to your lives, you will find yourselves vibrating at a higher frequency.

The key word here is 'applied'. There are many people on the planet who have a great intellectual understanding about energy but refuse to apply it personally. It is as it is. It takes courage to choose to be your own guinea pig but how else will you know for sure? Reading about it will only increase your intellectual knowledge.

Now to the first exercise.

Ask yourself whether you can understand and feel how Ray One works through you in your life. How do you use your free Will and your Power? Do you live your life making decisions for yourself, or do you wait until another makes them for you?

Should you be one who makes your own decisions, then you are using your free will to activate your power base within you. The opposite is also true. Should you wait for others to make your decisions for you, then you erode your personal power base. You give away your vital life infusing energy to another, and eventually your health will suffer.

Question self as to whether you force your Will upon another and expect them to act accordingly. Is this not an abuse of your will, your power? These then are samples of the themes raised in the understanding, and the application of, First Ray energy as expressed through the Aries sign.

As a probationer, the lesson is to work with this theme using your natal birth chart as your model. During the next few days there will be incidents arise in your life in which you will have the opportunity to observe yourself and your use of Ray One, the Ray of Will and Power. This is how energy works.

As you focus your consciousness on a thing that consciousness draws an experience to you. This is called the Law of Attraction. The Law of Attraction is one of the spiritual laws and principles of the Universe and is part of the Mystery School teachings. Consciousness is energy. Energy is thought. Whatever you focus your energy upon attracts like to you. Should you use your free will to focus your thoughts on lack, this is what you will attract.

Consider this deeply. Abundance and prosperity occur when your consciousness experiences abundance and prosperity. When your consciousness experiences lack, you will have lack.

You can create whatever you want for your life. You need only to focus your will in that direction. Should you choose to use your will to change your thought forms consciously, then change will be attracted to you. And, so it is.

Reflect on how many times a day you have the choice to use your will constructively. And then ask yourself the question "How many times a day do I use my will constructively?" You are the author and the creator of your own life. It is through Ray One that you can create magnificently.

Ray One, the Ray of Will and Power through Leo and Capricorn

When one has advanced sufficiently along the path to the light of conscious self-understanding and can feel and personally apply the cosmic energies constructively, a whole new state of being manifests. Worries and cares fade away. The initiate is now ready for a more advanced level of training.

This training can come in many ways. Most often a group of inner teachers prepare the initiate for their next evolutionary step by presenting them with life situations that require Right Use of Will. In other words, the initiates are tested to see if their will is strong and can carry them through to a higher frequency. The initiate of course may be unaware of this.

Using the will consciously, through the sign of Leo, can be quite challenging, as, by this time, the probationer is involved with group work. Often, the adulation of the group can bring such a strong surge of emotion that it becomes addictive. It takes Right Use of Will to remove oneself from these situations in order to focus on true purpose.

Downloading through Leo, the First Ray, when used constructively, brings the group leaders into more exposure by which they will be tested with this leadership situation.

Many on the spiritual path fall by the wayside at this point in their spiritual evolution as they begin to believe themselves to be great gurus or teachers. This illusion creates a dissipation of the love energy, and the personality self begins again to take over from the Soul. The outpouring of the First Ray energy through Leo is a major initiation point for the journey of the Soul.

Those who are group leaders among you question yourselves as to whether you allow the adoration of your group to feed you, and whether you allow yourself to become addicted. Do you so long for approval and recognition that you are unable to break yourself free from this? If this is the case, the by-product will eventually be an erosion of your power. Your power base comes from Right Use of Will.

So, we move on to Capricorn. How then is this First Ray energy experienced through this sign? How best can you use your free will to bring about true power?

Question yourselves as to whether you use your free will to control other people in order for you to have your way? Do you think this is an appropriate use of your free will? Maybe you use your will to experience paralysing fear and allow that to restrict you?

There are of course many levels, and there is a level through which you will be tested. Become aware of when this occurs, and commit to moving beyond it. If you allow the restriction, your soul force begins to atrophy and can remain in that bind.

As you can see, the probationary period of three months requires great endeavours, much awareness and the willingness to move on. Many people find this phase the most difficult. It is as it is. There is always plenty of time, another Capricorn theme! The journey of soul into union with The One goes on and on. There are many, many levels.

Beings on Earth, you are so powerful and through your Right Use of Will you can demonstrate your power in a way that serves the greater good. It is through the sign of Capricorn that initiates can structure their lives so that time is allotted each day for quiet meditation and reflection.

Discipline is required. This conscious use of Will on a daily basis enables the soul to be in contact with inner guides and teachers from higher dimensional planes. It is through this regular practice that much training takes place. Ask the author. She used her will in this direction resulting in clairaudience, through which these teachings can be made known to the world.

In all things in life one has a choice, a choice in how to use one's free will. Your life will reflect the demonstration of your choices. Should you use your Will in such a way to allow fear to rule then your life will be a demonstration of that. The Law of Attraction always works. Whatever you fear, you attract. Is it not more rewarding to work through your fear and face it courageously through Right Use of Will, rather than create lack and limitation in your life?

All probationers will be faced with a fear during their three months' trial period. Fear is illusionary. The thought behind the fear is what is so limiting. Become aware of the thought and using your free will, focus your consciousness, on the result of moving through the fear. Allow that to guide you. Fear is a distraction, and is simply a test to pass to enable you to move on in your evolutionary growth.

You can see now how Ray One, the Ray of Will and Power works through these aforementioned three signs. Once the initiate has mastered the Right Use of Will through these three signs, the urge to use the Will in service to The One becomes of paramount importance to the life. The way this will be done is through that Soul's own creative impulses and the love of that work. Discipline and structure are required more at this level.

Often the First Ray energy brings with it the sequence of Capricorn, Leo and Aries. Firstly, tests are necessary to move through the fear, then the creative urges are felt, and then the initiatory action is required. This is often the process. The consciousness required for this process and the resulting Right Use of Will can be highly challenging. It is a choice.

*

During the Mystery School training, initiates and disciples (those further along the Path) were given many tests by their inner guides. At the same time, they received teachings from their physical teachers. And, so it is today. It is up to each individual on the Path to become aware of these inner tests.

Mischievous entities can enter your psychic field from time to time, sometimes at the request of your inner teachers, in order for you to learn discrimination and discernment. It is all part of the training. Many aspirants become mesmerized by these mischievous entities, and the games they play, especially when they appeal to the ego. The author can give you many examples of this.

Discrimination and discernment come under the influence of the sign of Virgo. In order for perfection to be experienced within the mind, body and soul, discrimination and discernment on these three levels need to be experienced.

Discrimination of the mind involves using your free Will to choose which thoughts to think. Discrimination of the body involves becoming aware of each nuance in the body and recognizing it for what it is. Any discomfort, pain, spasm, restriction or weakness in the body is simply energy behaving dysfunctionally. All cells in the body contain memory in their etheric field. When the memory is triggered by a word, thought, or emotion the cell re-acts according to that memory.

Discrimination and discernment are needed to develop an understanding of the energy blockage in order to clear it at that time. As probationers on the journey to the Light, you are required to develop this awareness. Your Master Jesus, the most highly evolved being to have physicality on your planet, taught this truth via his parables. Another will come to teach you in much the same way.

When one is fully committed to the Path of the Spiritual Disciple, one learns, through trial and error, how to come to a place of joy within. Recognition and acceptance of one's own journey, and the understanding of the process, engenders this joy. Internal peace is experienced when discrimination and discernment are learned and fully integrated. Surely joy and peace are worth the effort!

We on the higher planes of existence admire you who chose to take on physical form. It is a mammoth task you have undertaken. The outcome is always clear. Each soul will return to The One. How long it takes depends on each individual's free will choice.

There is no such thing as time in our dimensions, so the concept is irrelevant to us. It is only on the Earth plane that time, a Capricorn theme, seems to have importance to you. Within the greater scheme of things time is not matter! It does not matter.

Let's talk about Love for a moment. Love does matter. Love is the highest vibration of energy and it is only through consciously and daily developing the feeling of love in your hearts that soul acceleration to the Path of Unity can take place. Love takes many forms, and this author has written a book on these different forms.

Ultimately, Love is a feeling that is personally experienced when one totally honours, respects and values oneself for who one really is. Each human is a part of The One, Great Spirit or God/Goddess, whatever is your terminology. You each have that spark within you, and it is without you in everything that you see, feel, touch and hear. It is. You are that.

When this Love is totally accepted and felt internally as a burning flame, your vibration changes and you become lighter. You exude Love and lightness and thus attract it to you. This is the universal Law of Attraction at work.

Ray Two of Love and Wisdom through Gemini, Virgo and Pisces

On your Earth plane, the Master Jesus, known to some as Yeshua, was the manifestation of Christ energy. The Buddha was the manifestation of the Wisdom energy stream of consciousness. These two beings of high evolvement chose to enter your third dimensional realm in order to anchor the Ray Two vibrations on to your planet at the time they did.

Loosely speaking, this was during the Age of Pisces. In the Aquarian Age the activation of these vibrations is awakening in human consciousness. In order to use these qualities and become them, you need to experience how they work. This is the only way to know. From experience comes knowledge, and from knowledge comes Wisdom. From Wisdom comes Love.

Let us talk about systems. There is an order to creation. Our Great Source, All That Is, created a system of energy for all beings to utilise. The One created free will to enable all of its creation to use it to work the system. Esoteric Astrology is the tool to do so.

Let us encapsulate this lesson.

First, there was the One. This One is the Supreme Creator of everything. Our Prime Source initiated action on an idea. The idea was to experience itself through creating many different aspects of self. A system was needed to enable this to be made manifest. Distribution of energy was the system. All of creation is energy. All is connected by energy.

Great Rays of energy were created that were stepped down to less intense frequencies. Much stepping down was needed. You could say that power stations were created with many sub-stations, distributing power to many areas of the cosmos. This energy system can be understood, through experiencing. Intuitive esoteric astrology will give you the information.

As you know, communication in your world is so vital. The communication styles you use are many and varied however, only infrequently are you able to communicate as we on the higher planes do, that is telepathically. In times to come, humans will again communicate this way.

As always with energy transference, it is the intent behind the energy that is the key to the kind of energy that is received and transmitted. When communicating verbally or through the written word, if the intent is to do so lovingly, that intent carries the frequency of love.

If, on the other hand, the intent is that of anger, frustration, jealousy, revenge or any other base emotion, then it is that energy that is received and transmitted. As a probationer, your role is to become consciously self-aware of the energy behind your communication, and to question yourself as to whether it contains the vibration of love.

At the Mystery Schools, probationers were tested many times a day, often unbeknown to them. They were always taught about the importance of communication yet often, through excitement or agitation, they would forget themselves and go back to their old ways.

As mentioned previously, wisdom comes from knowledge gained through experience. It cannot be otherwise. Love is a vibration that is experienced within and comes from the heart. Love cannot be experienced within the heart when another base emotion is being expressed, especially that of fear. Therefore, to actively and consciously utilise Ray Two, The Ray of Love and Wisdom, one needs to have consciously worked through identifying and clearing these base emotions, in order to access the love vibration that is within every human. This comes about by developing wisdom through self-knowledge.

When you totally accept, honour, respect and love absolutely your whole beingness with all its many facets, then you can begin to work consciously with the Ray of Love and Wisdom. Its strength will develop more forcefully within you when you communicate your truth from your heart, with love. As this occurs, a different vibratory quality develops within the voice, and a purer sound is the result. The loving vibration emanating from this sound is transmitted to others.

All Masters know and understand the extreme importance of this use of energy. Masters cannot become so until their whole being vibrates to this frequency. The role of the probationer is to consciously develop a higher frequency of thought and speech by always communicating truthfully from the heart and choosing to master one's thought processes.

Your planet is ruled by Ray Two, the Ray of Love and Wisdom, or the Christ and Buddha consciousness streams. We on the higher planes, do not see you as form. Your thought-forms create your frequency When your thoughts are totally focused on your material plane pursuits your frequency is lowered, your light is dull and we cannot access you.

When each one is willing to focus attention on one's thoughts and feelings and willingly change those that are self-destructive, a lighter frequency is experienced, and seen by those with eyes to see. It is then that you begin to access Ray Two. This energy is down-stepped through Virgo, Gemini and Pisces, and is always accessible. Your planet is bathed in its frequency.

The Love and Wisdom emanating from this great being of Light is so vast it is almost beyond human understanding. All humans on Earth will return to union with The One. This Ray will hold steady its frequency. The timing for each of you to access this potent energy stream is your choice. Many are impatient and yearn for your return home, yet are unwilling to make the necessary preparation.

Is it not so that when you decide to take a journey, preparation and planning are needed? So it is, with your journey home to wholeness and unity with The One. The preparation and effort involved brings you to the role of the initiate, and opens the door to become a disciple.

There are many levels of existence. It is on the Earth plane that one experiences emotions and sensations. These can become addictive. Addictions come in many forms. There can be addictions to food, alcohol, drugs or sex.

There can also be addictive behavioural patterns, and addictive emotional re-actions. As an example, it can feel temporarily empowering to dump your aggression onto another. It is through the sign of Pisces that these addictions can be understood. Love and Wisdom cannot be expressed if blocked by addictive thought and emotion.

It is through loving compassion to yourself that you can break through addictive patterns. Energy just is. It is what each individual chooses to do with the energy that determines the results in one's life. Part of your training is to learn to identify your addictions and then do the inner work needed to consciously release them.

The strength of your desire to move forward on the Path will be the determining factor. When you use the qualities of Virgo, those of discrimination and discernment, to concentrate on, analyse, accept and heal the addictive traits, much progress can be made.

Jesus embodied the Christ energy on planet Earth in the Piscean Age. He exuded loving compassion to all. He did not make distinctions between rich and poor, literate or illiterate. He regarded all equally. He knew and understood that each individual was God's creation and was connected by a stream of energy.

He treated every living thing with respect, honour and love. His frequency was such that all felt great love simply by being in his presence. He knew he was part of God and knew that everything in creation was also, therefore how could there be any division? This is great wisdom and truth.

Jesus demonstrated great faith. He had learned to experience and to feel the vibration of faith coursing through him and asked his followers to experience it also. Faith, an attribute of Pisces, can only be developed from within. It comes from learning to trust the essence of love within your being.

In the ancient Mystery Schools, much focus was directed towards personal growth through experience. Lessons were taught through play and creativity. When one learns this way, the knowledge taught is readily re-accessed when needed. So many people in today's world have difficult experiences learning the current education curriculum, so the information given is blocked. Schools in the future will understand this, and will encourage teachers to pass on their knowledge through creative, playful pursuits.

One way you can learn the content of this book is by playing with it daily. Become aware of the psychological games you play and choose to create games that are more fun. You will find your life will become lighter and much more enjoyable.

Let's look at the second Ray of Love and Wisdom again in the light of play. Ask yourself the question: Do I play with my work and have fun with it, or do I work in order to play at weekends? Do I enjoy the work I do? Does my work lift my spirits to great heights, or does it deaden my spirit?

When we spend our days loving ourselves enough to work with what we love, then we pass this frequency on to others.

Each human has unique gifts and talents and this can be understood and developed through following the dictates of the heart, rather than the head. The head is so full of old psychological programs and patterns from others such as parents, family and society. These programmes belong to others. They are not your own.

In order to live a life of joy and happiness, filled with love, you need to follow your inspiration and joy. Eventually it will lead to your soul's purpose.

Ray Two, the Ray of Love and Wisdom, affects every living thing on planet Earth. The great Lord of the Ray lovingly holds it in place so that, eventually, all beings on Earth will emanate this frequency to others.

This is the Great Plan for humanity. Each of you can make your own unique contribution to The Great Plan by consciously choosing to live your lives full of love and joy.

Wisdom develops by examining the many varied experiences created by your thought patterns. Wisdom develops through experience. Have you ever said to yourself "I am never, going to do that again!" A life lesson has been well learned and integrated, through experience. You chose to enter the Earth plane to experience, and as you do, Prime Creator, Great Spirit of All That Is, also experiences, which was the original idea.

As we are all a part of God, each having a divine spark within, we are all connected to each other. When we feel love, joy and compassion, we feel more connected to others. Love and joy come from following the path of the heart, the path of the Soul. Wisdom comes from the many and varied experiences along the way. And, so it is".

Ray Three of Active Intelligence through Cancer, Libra and Capricorn

"What is intelligence? Does it come from intellectual study of many and varied subjects, or does it come from innate understanding and inner knowledge? Intellect and intelligence are as diverse as oil and water. Intellect requires the concentration function of the left-brain, the logical, rational and analytical part of the brain. Intelligence comes through the use of the spatial, conceptual and intuitive hemisphere of the brain.

Active Intelligence arises from the balanced use of both modalities with the left-brain, the logical rational mind, following the direction and guidance of the intuitive brain. The intuitive brain connects to the soul and the heart, and carries the soul's wisdom and truth.

The left hemisphere of the brain comprises many compartments of learned and conditioned information that may or may not be true and useful to the soul. As a probationer on the Path to the Light of unity with The One, you need to learn how each of the two brain functions work.

Most people in your western world have allowed their intuitive, conceptual and abstract brain to atrophy, focusing instead on purely logical, rational thought forms. The soul does not understand this. It is spatial, and understands images and symbols, and is always able to see the big picture.

The personality ego-self, on the other hand, can only understand the little picture, the one it is immediately focusing upon. The soul then feels alienated, and the individual will eventually feel unfulfilled and incomplete.

The probationer is encouraged to allow, and to trust, the impressions and impulses that first enter the mind. Then, through heightened awareness adopt them and act upon them. The soul provides information at a faster speed than a modern computer. The personality self needs to become conscious of these impulses.

The path of unity, wholeness and love requires following the lead of the soul, and the soul leads through the intuitive process. Soul contains the knowledge and wisdom of experiences gained over all its many and varied incarnations. How could the personality, which is a one-time composite of the soul, possibly learn through book knowledge all that has transpired during those times?

The only way this knowledge and wisdom can be obtained is through tapping into the Soul and allowing it to speak to you. You can do this in a variety of ways. Listening, or paying careful attention to your body is one way. Each cell in the body contains Soul memory. When the soul experiences discomfort, it will make its presence felt by expressing that discomfort in the body. It may manifest as a pain, itch, heat, cold, spasm or illness. It is one way of bringing attention to its discomfort or pain.

Another way is to learn to decipher and interpret your dreams. Dreams contain symbolic messages direct from the soul. It takes willingness, effort and discipline to learn to read your soul's messages.

A third way is to look for the hidden meanings in the omens and synchronicities that happen to you daily. When you become aware of their symbology, you will soon see the 'cosmic joke'. Life is a game. It can be a fun game, or otherwise. You choose.

Our channel has learned to access Ray Three so is able to write valuable books by using her intuition. She accesses her Soul's memory banks, takes notice of her dreams, and uses the information to play her game of life.

Able to engage her left hemisphere function to analyse the dream, she then allows her intuitional insights to interpret truth. She feels her soul, and knows when she has interpreted correctly. She uses this information to guide her in her life. She has learned to pay attention to her body, becoming aware of each area of discomfort. She talks to her body to comfort it.

As an example, she has many, many soul memories of fearful situations stored in her knees. Her knees begin to pain when she is being urged from within to take the next forward step on her soul's journey to the Light. She understands her body's messages and directs her full attention to her knees. She comforts them, assuring them she will move forward at a pace that is steady and cautious. The pain then subsides because the soul memory is pacified.

Active Intelligence comes from allowing the soul to express itself through the intuitive process, and the sensitivities of the body. The Third Ray of Active Intelligence is stepped down through the signs of Cancer, Capricorn and Libra.

When you allow the intuitive process to guide, and when you become consciously aware of the process, the soul's path can unfold. The soul develops a sense of belonging, a Cancer theme, along with a feeling of coming home. It begins to feel secure, and safe within its body home.

The personality-self or ego begins to relinquish control, a Capricorn trait, and fear subsides. Warmth takes over from coldness, and the two opposing signs of Cancer and Capricorn come into balance and harmony through Libra. It is so. It is simple.

As a probationer, are you paying full attention to your body's many messages? Do you interpret your dreams from a symbolic perspective? Do you allow their messages to guide you in your life? Are you aware of the omens and synchronicities that occur daily for you, and do you consciously act upon their messages?

Your soul tries to get your attention. Do you play your part in the game of life by being attentive to these messages, acting upon the soul's wisdom and guidance? Or, is your conditioned left brain far too busy?

Of course, the path of the initiate and disciple requires discipline. In order to activate the Third Ray much discipline is required. Time needs to be allocated daily to pursuits such as meditation that will release the hold of the rational brain. Time also needs to be spent on re-structuring mind-programmes, that create constraints. A commitment needs to be made to work with the soul. Dedication is required if you are serious about taking the journey to the mountain-top. You will need to become sure footed, mastering each step along the way.

Balance and harmony can be achieved when the personality allows the soul to lead. Practice, patience, discipline and a serious approach over a period of time need to be expressed. The soul rejoices through the process, and when results are attained, so it is not arduous. The time this takes is but a blink of an eye in a lifetime.

When it is time for the soul to accelerate its journey home to Source much emotionality can be experienced. There is a battle as the ego self does not want to relinquish control. Old dense emotional energy needs to be cleared so the soul light can shine. It is at this point that full attention needs to be given to the process so that the inherent intelligence can emerge from under the dense emotions. Should you use Capricorn control to block the emotions, unwilling to feel their truth, active intelligence remains suppressed.

The journey to the Light requires active participation and self-awareness by the individual. It requires the use of free will, the desire to feel love and develop wisdom, and the ability to activate and use intelligence.

As probationers, are you dedicated enough, committed enough to follow this path? Do you allow your intuitive perceptions to guide your actions? Do you pay attention to the omens and synchronicities in your life? Do you take notice of, and utilise your dream teachers' messages?

The Ray of Active Intelligence is available to be used. The great Lord holds it in place so that all humans can access it. The result of its conscious utilisation is peace and harmony, and the ability to see beauty in all things and all people. The Path is clear and the road straight. The journey home is now under the leadership of the Soul. And, so it is.

Ray Four of Harmony through Conflict through Taurus, Scorpio and Sagittarius

How many humans have experienced conflict? It is unnecessary to live one's life this way, however it does seem to be the human way.

When conflict is experienced outside of self, it is purely a reflection of internal conflict. Whenever you create disharmony in your external life, it is the result of inner disharmony. You may not be consciously aware of this inner turbulence until the disharmony externally is experienced.

Energy simply is. It exists to be used. How it is used depends on the consciousness and will of the individual. The great Lords of the Rays beam down their energy to you, unceasingly. What you do with it is your free will choice.

The path of the disciple is to learn to feel the differing qualities of the cosmic energies as they express through the zodiacal signs, planets and Rays. Once felt, the task is to consciously and constructively use them. It is through this conscious use of cosmic energy that conflict will be dissipated and harmony takes its place.

There can be a great internal battle take place because the ego will go to great lengths to have things remain the same. It takes conscious effort and discernment to become aware of this process, and to choose to work purposefully towards the light. The path can be long, and many are left by the wayside, attracting more conflict within and without.

How you feel about your life is but a reflection of how you feel about yourself. As an example, should you feel hurt and inferior to a sibling then that hurt and inferiority covers a memory. The memory contains energy that you used destructively at some time. Conflict will always be experienced until such time as a conscious effort is made to deal appropriately with the stuck energy. It needs to be freed so that it can once again course freely through the body.

When disharmony is experienced, go to the place in the body where it is stored. This requires dedication, time, commitment and willingness. One needs to quiet the mind, relax into a meditative state, and ask the body to reveal the area where the conflict is stored, and the issue behind the conflict.

You will receive a picture, sensation, word or impression. The body will respond. Go into that place to access more information. Once the issue is understood and integrated, breathe deeply into the discordant energy. The extra oxygen will assist the blocked energy move. Consciously guide the energy up your body and expel it through your mouth. It may not taste pleasant, and it may have a foul odour. It has been a resident for a long time. It needs to be expelled from your body.

When it has released, imagine it transmuting into light. Then imagine the space where it used to be filling with light. By using this technique, you are consciously bringing in new energy to replace the old. You will feel the difference and will find that any future interaction with the sibling will no longer upset you.

Should the dense energy remain and you choose not to make a conscious effort to deal with it, the conflict between you and your sibling is likely to escalate until the energy finally bursts free.

As an initiate, you will have learned much of the above. You will have learned to recognize internal conflict and will understand how to use energy constructively. There are many layers.

Imagine your body as an empty form. Inside that emptiness are a great number of sparks moving in all directions. Some sparks seem incredibly bright, and some are dull. There are likely to be spaces inside your body where there are no sparks. Your fire is very much dulled in these areas.

As you are aware, it is the transiting planets moving through the constellations that activate many of the dull areas of your inner space. This is the time your soul has chosen to experience a stirring of a particular energy. Each planet and constellation have a different characteristic to its energy, and the characteristics contain certain themes.

When a beam of energy enters an internal dull space, activation takes place. This energetic activation stirs soul memory. It stirs up emotions. E- motion is energy in motion. You, as probationers on the Path to the Light, have the challenge of dealing with this energy appropriately.

Firstly, you need to become aware of the emotion, identify it and own it as yours, "I am feeling ... inadequate and inferior", for example. Acknowledge that it is this emotion that is causing you to feel internal conflict and disharmony. This conflict may or may not have been triggered by another.

If so, that other is rendering you a tremendous service, as they are the catalyst for your growth to conscious understanding. Words, whether verbal or written, are triggers that activate energies. Words are so powerful. Thoughts are words strung together into a sentence. How powerful are thoughts. Much internal conflict can be experienced, simply through thinking.

In order to achieve internal harmony, one's thoughts need to be mastered so that any thoughts of conflict can be consciously released. This requires the willingness to slow down one's life sufficiently to enable the implementation of this process. Training the mind to think only constructive and harmonious thoughts is part of the probationers' training program.

Daily meditation is one tool to accelerate this process. Deep conscious focused breathing is another. When combined, the thought patterns begin to slow down, and the probationer can then observe them in a detached way. This constant practice leads to mastery of the mind. Of course, discipline is required, as is the desire to move beyond conflict.

The three signs through which this Fourth Ray energy is expressed are Taurus, Scorpio and Sagittarius. As Taurus energy can be slow to move and hard to budge, the conflict experienced can be great. Those dark places within the empty spaces of the body are likely to experience density. Many deep and unconscious blockages of energy can surround Scorpio. These two opposing signs, when activated, can bring great internal conflict, leading to external conflict.

When energy is blocked, unable to move freely because of repressed emotions, Taurus, through insecurity, lack of self-worth and self-value, will stubbornly remain in that same psychological place. The probationer will experience this, and the unconscious issues may manifest through money problems. Conflict occurs also through Taurus when one is greedy, holding on to money through fear of lack. How can abundance, and the state of harmony, exist in this atmosphere of possessiveness?

How much conflict does the unconscious misuse of Scorpio energy create? Do you see the results of this abuse in your world today? As a probationer, do you play manipulative games over others? Maybe ingratiating yourself with another in order to gain benefit for yourself. Maybe you exert power over another by covert manipulation. Maybe, through envy and jealousy, you give away your power. Maybe you harbour deep resentment. Do you not agree that these states of being result in conflict?

How determined are you to create harmony in your life? Are you willing to do the inner work involved to correct your abuse of energy?

Maybe you preach your gospel of beliefs forcing them upon others? Have you experienced your beliefs personally, or are they book knowledge?

A person's beliefs are true for them. They may not be true for another. Look around your world and see how much conflict there is by the abuse of this Sagittarian energy. How many cultures have fought religiously based battles? How many millions have died through the enforcement of religious doctrine?

How much more harmony there would be if each individual learned to develop their personal spiritual truth and only pass it on when asked to do so? How many people give their power away to religious and spiritual gurus or some entities that pose as guides?

External harmony can only be experienced when you have learned to create internal harmony. This Fourth Ray energy manifests through two fixed and one mutable sign. Inner conflict can only be resolved by your willingness and desire to do so.

By owning your own conflict, and not blaming it on another, you begin the process of creating harmony. Once you own, and identify, the dysfunctional use of the energy, you can begin to use it consciously by allowing your intuition to empower you. You will utilise the positive qualities of Taurus, those of persistence, stability, consistency, perseverance and determination. You will become aware of the beliefs that dis-empower, and learn to access truth from within, and live it. Harmony will be the result.

Ray Five of Concrete Science or Knowledge through Leo, Sagittarius and Aquarius

In the beginning of our story, we mentioned Machu Picchu, high in the Andean Mountains. This sacred place contains memories of the past. People who visit there can feel these memories awaken within them. Should these people so choose, these memories can be accessed.

During Atlantean times great spiritual progress was made, and many highly evolved beings taught at this Mystery School. At that time, women were regarded equally to men, and were highly respected for their intuitive abilities. No decision was made unless female oracles were consulted. This level of intuitive wisdom existing at that time will again become prevalent on your planet Earth. It is time.

As higher frequency energies bombard your planet and Solar System, your higher intuitive brain centres are stimulated. It is through this stimulation that greater intuitive and oracular abilities will manifest. The time of dormancy is over.

Even now, many governments around the world employ intuitive people to aid their highly secretive research programs. They like to keep this activity secret. Many humans are discovering intuitive diagnostic medical abilities and mainstream medicine is starting to take notice. Police departments are employing intuitives to assist with crime solving. The list goes on. This process is accelerating and once again women will be respected for these natural abilities.

These are the type of ancient memories taught and practiced at Machu Picchu. The stones and crystals contain information. Knowledge is stored in 'concrete', in rock. When you visit this sacred place, do so with greater awareness.

Allow the memories to surface. Allow the abilities, well learned during that time, to come forward. These abilities may need re-training, as they have been lying dormant within your cellular memory for a long time.

This knowledge is the 'Concrete Science" of which we speak. It is only through the intuitive process that this source of energy can be experienced, and thus understood. When understood and integrated, it becomes concrete so to speak. It remains in the memory until re-activated by specific frequencies. This is what is happening now.

Ancient concrete knowledge will once again be known however it will appear new to the masses. It may be regarded suspiciously until proven otherwise. Many sensitive men are uncovering these abilities. Memories are stirring. Some understand, and are learning to trust, whilst others fear the memories and attempt to block them with alcohol, sex or drugs. The memories are mostly unconscious, at first.

The increasing force of photonic energy can create mental health problems, unless understood and used constructively. It takes trust. Trust that your body does indeed know truth. Trust that the knowledge is relevant and is in fact 'concrete'.

As the precession into the Aquarian Age accelerates, the higher energy centres in the human brain will be stimulated. Knowing this energy exists will assist its management.

Aquarian energy, electrical in nature, can be experienced internally, somewhat like a lightning strike. A sudden awakening or realisation, can occur and an atrophied part of the brain begins to regenerate. Excitement is experienced, and one begins to view life differently.

Ancient wisdom and knowledge, through Sagittarius, awakens. More insights and visions are experienced and intuitive abilities enhanced. Many will not believe. Their safe and known world is collapsing. Many established systems will break down, and this may cause fear.

It is optional how you use the Ray Five energy of Concrete Science and Knowledge. Many are now accessing it and we are assisting. Many people awaken in the early morning with brilliant ideas they know will work, and are able to apply them to their current projects.

When one is operating from the heart centre, living life doing what one loves to do, being totally true to self, in other words utilising Leo and Sagittarius positively, then one becomes an open receptor for the Ray Five energy through Aquarius. This energy exists. One needs to become conscious of it to appreciate it working.

Your planet is on the threshold of revolutionary change. Many people are experiencing huge shifts in consciousness. Some have the courage to be different, to stand in their own truth and be counted, no matter how crazy it may seem to others. They know they face ridicule however their concrete knowledge and inner wisdom drives them on. They have learned well.

Science, in ancient days, was different from your science of today. Science then was the study of Nature and the energies of Nature in all its many and varied forms. To understand the cosmic forces again, this ancient science will once again come into prominence, however many structured mindsets will need to be struck by lightning first!

As with the zodiacal energies so it is with the Rays. Each individual experiences the cosmic energies within, to differing degrees. Most of you reading this book will be working with Ray Five, the Ray of Concrete Science and Knowledge.

Those not allowing yourselves to work at what you love to do, please take stock. It is only through accessing one's heart centre, and following its dictates, that the intuitive knowledge and ancient science can once again be made concrete within you. Should you deny love, energetic barriers form around the heart, blocking the love and creative flow.

When the heart sings with joy, the effect is greater warmth and openness. When the heart centre is open, higher energies can flow into it. As probationers, the task of heart opening may seem trivial and difficult to accept logically. Logic has nothing to do with it.

The heart is the power source of your body. Releasing all that blocks its efficient functioning is of paramount importance. So, probationers, we encourage you to play, have fun, laugh - especially laugh, dance, and become aware of the activities that bring you joy. Become aware of what makes you feel alive, happy and joyful, and question what is it that has brought you to this state. Become aware of when you feel excitement, and have the urge to do.

Follow this excitement. Allow your joy to be your guide. From this practice, your soul's purpose will begin to unfold. The more in love you become with yourself and life, the more openness you will feel in your heart. As the heart opens, barriers dissolve.

Intuitive abilities are connected to the heart. As these abilities develop, your heart feels joy. Encourage the process. Your soul's memories of natural gifts and talents will continue to unfold as concrete knowledge and ancient sciences begin to make themselves known. You will wonder how you know.

What an exciting life you can create for yourself, simply by following the Path of Love. When in this state of love, the love you feel within is experienced by all within your field of influence. Love is the highest vibration. It is through love that Ray Five can be accessed.

It is so.

Ray Six of Idealism and Devotion through Virgo, Sagittarius and Pisces

As evolving humans, many changes are to take place within the consciousness of the individual, and the collective. This is both necessary and vital. Unevolved humans have begun a journey to the Light, a journey to the merging with the higher aspect of self under the direction of the soul. The journey culminates in union with the Divine.

There are many steps along the way and many roads to take. Most humans take the bypasses in the hope that they may quickly lead to the end. The journey takes as long as it takes.

It is the journey that is important, and all that is learned and experienced along the way. As sparks of The One, as each human has an experience, so too does the Creator. This is the Great Plan, and the Plan is carried out.

Every human operates within his or her sphere of evolvement. At this present time on your planet, the masses are focused on the material plane. Money is their god, and spiritual values play a minor role. This is changing, as it must. The pendulum has begun to swing the other way. This swing will gather more momentum as you move further into the Age of Aquarius.

The path of the initiate and disciple is firstly to set one's ideal, something larger than self and that is for the greater good of humanity. When one is committed to this ideal, the sixth Ray can assist. Until you set your ideal, the Ray remains dormant.

It was a momentous day in the author's life when she formulated her ideal. As a student of astrology, you too can set a similar ideal and that is to reach the potential inherent in your birth chart.

Understand that each of you has chosen your birth time, place and environment. You so choose in order to evolve into greater Light and Love. Your birth chart is your guiding light – your energy map for self-discovery. All you need do is learn the system, and choose to use each of the energies constructively.

Devotion to your evolutionary growth is required, as is devotion to the path of unity with The One. Your birth chart is a model to show you the way.

When you work with your chart, identifying the specific energies being activated on a daily basis, you have access to the path of evolution and soul infusion. You have the choice to treat the information as purely intellectual, or be willing to use it constructively to assist your evolutionary growth.

Energy is. What each individual does with it is what matters in their evolutionary journey.

Let us return to the ideal. When formulating it, you need to use your Virgo analytical and discriminatory abilities to observe what it is that will bring you the greatest soul fulfillment. This will not be found in material things. Material things can bring temporary comfort, but ultimately are not fulfilling.

Our channel had no idea of her direction or path. All she knew was that she wanted to understand herself on a deeper level, so she could move away from unhappiness and depression. She had a burning desire to perfect that which she allowed to manifest as imperfect within her. She understood that all the answers to her difficulties in life were shown in her natal chart. Her desire to move beyond disharmony and conflict urged her on.

All emotional and mental discomfort, difficulties, depression, hurt and weakness are caused through misuse of energy. All disease is simply blocked energy. It is by consciously using the positive expression of the zodiacal signs, the planets and the Rays that ease and harmony can be attained. Within the space of a life, the time needed for this journey is miniscule. It depends on how devoted to your ideal you are.

The Virgin, in ancient times, was regarded as a female who was whole and complete within herself. She had no need of another to feed her energy. She was fully contained. From that state of being she freely gave herself in service to others. She was highly regarded by the masses as someone who had attained a high state of evolvement.

She was able to use her deep wisdom through Sagittarius and compassion through Pisces, to serve and heal others. On the completion of her daily work she would seek solitude to replenish her energies. She knew she could not continually give out her energy resources to rescue and save others. She was able to discriminate wisely. She knew her health would suffer if she gave too much.

This state of being was demonstrated and experienced by the women resident at the Machu Picchu Mystery School. The teachers were healers, and the healers, teachers. All of them operated individually, utilising their personal energies constructively to aid the whole.

Probationers, you too can achieve this state of wholeness, wisdom and compassion simply by focusing each day on feeling into the energies that play inside you. Using your Virgo discrimination, choose to move up the evolutionary ladder to unity with The One. You will need to study each sign as its energy manifests within you, and choose to use it positively. You will be placed in many situations where you can put this knowledge into practice.

Pisces brings the energy of Divine Love. This is a higher, and more refined love than that of Leo. It requires detachment from the being upon whom the love is focused, whether that is self or another. It requires an objective awareness of self, feeling and experiencing. It does not require a dependency on the other returning love.

When you contemplate Divine Love reflect on how the Master Jesus expressed it. He was not attached emotionally to his family, his disciples or his followers. He loved them unconditionally and accepted them exactly as they were. He gave healings and teachings whenever he could.

He was of such a high vibration he did not need a great deal of replenishment time however he did spend much time in nature, meditating and connecting with The One within. He personified Virgo wholeness, Pisces compassion and pure love. He had an ideal, to do his Father's work, and he fulfilled his evolutionary challenge. He was devoted to his ideal, and nothing would deter him. He said that whatever he did, so too could you. He fulfilled the potential of his birth chart, his purpose for incarnation and left a legacy for all humans to follow.

There are many humans now activating their internal Christ energy. It is within everyone. Accessing it takes application and devotion to your evolutionary ideal. The ideal needs to be broad in scope - Sagittarius. Your soul knows the big picture of your incarnational purpose, and will urge you to follow this path. It can only enter your consciousness through feeling and intuition.

The challenge, probationers, is to formulate your ideal. Be discerning. Why not use the exoteric ruler of Virgo, Mercury, to write it? Allow your intuition and inner wisdom to guide you. Prepare your Sagittarian bow and let the arrow fly high. Be subjectively analytical through your chart, and choose to use the energies constructively and positively. Practice devotion to your ideal and thus access and use wisely Ray Six.

Ray Seven of Ceremonial Order through Aries, Cancer and Capricorn.

And now, probationers, we are at the last section connected to the Rays and the zodiacal signs through which they beam their energies. Much has been learned, studied and applied has it not? What is to be uncovered now?

People experience Ray Seven, the Ray of Ceremonial Order, in differing ways. The sign of Capricorn brings the theme, that of organization. As you are aware, there is a great order to the Universe. Your Solar System is a precise and calculable system. It is planned and orchestrated with precision, and has been organized to function in a specific way. Its movement is synonymous to human evolution.

What kind of mind is needed to produce such a plan? Can a human possibly conceive the enormity of the task of creation? There are many light beings administering the order. They manage the system well. The beings from Sirius have the responsibility of holding the orbits in form. They are masters at assisting the Lord of this Ray.

The ability to organize and create wonderful ceremonies, from a spiritual perspective, is developing well in your world now. Many great ceremonies are now taking place regularly upon your planet. The Harmonic Convergence was one great one that instigated light. Anyone who works with the organization of spiritual events is working under this great Ray.

Firstly, one needs to engage the energy of the Aries sign. An idea is presented to the mind and the appropriate action needs to be taken. Planning and organization are needed. A structure is also needed. One needs to responsibly assert authority. Sensitivities to the purpose need to be kept in mind at all times, knowing it will nurture many. Focus and mental acuity are needed.

In ancient days, attention was given to creating ceremonies for special events at the solstices and equinoxes. It was known that at these times a potent energetic window could accelerate the evolutionary path of the aspirant, should they participate. And, so it is today.

These events need to be carefully planned and lovingly executed, always with the end result in mind. This is to provide the participants an opportunity to move further up the spiritual mountain. Can you observe the use of Capricorn key words and themes?

When a disciple feels the inner urge to begin the task of planning a spiritual even, you can know they are being contacted by Ray Seven. The tests do not become smaller. As each event is successfully completed, a bigger one is presented. The reward for doing a job well is a bigger job!

It is always the free will of the disciple as to whether they continue under this Ray. Events need to be nurtured, and the participant's feelings need to be considered. The goal is harmony. If harmony is not present the attendees will not be able to access the window of opportunity presented to them.

Spiritual ceremonies are growing in number on your planet. We use the term 'spiritual' in a different way from religion. Religion is a doctrine and usually implies control. Spiritual implies no doctrine yet allows individuals to develop their own beliefs, through experiencing.

With any new Astrological Age there are always new beliefs that sweep the planet. It is a clearing out of the old and a bringing in of the new. One needs to be discerning, and sensitive to these new concepts and philosophies. One needs to experience for oneself these seemingly new ideas. One needs to stand above the many to feel how the new fits. One needs to take responsibility for how one feels, and be sensitive to the soul force within. Question self: "Does this ceremony stir warm memories within me?"

As always, there will be false gurus purporting to be great masters. It is only through the discernment of the heart and soul that one can feel truth. Our channel has had many tests along these lines.

Gurus come in many forms and in many modalities. It is up to each aspirant, initiate or disciple to feel into self to arrive at one's truth. When the initiates have demonstrated their ability to discern the true from the false, and when they have sufficiently demonstrated their willingness to allow their sensitivities to guide, they begin to work under this Ray of Ceremonial Order. The Lord of this Ray continually beams his great energy to your Solar System, and to Earth via the three zodiacal signs of Aries, Cancer and Capricorn; signs of the cardinal cross.

By the time the disciple is ready to work with this Ray, they have begun to mount this Cross and their duty and responsibility to humankind begins to manifest. Their soul's purpose for incarnation can be made clear, and their inherent expertise put into practice. Misuse of this Ray will eventually bring about the disciple's fall, so one needs to act with caution. Are you aware of the key words and themes of the signs being expressed?

As probationers, you have this role to look forward to. The training period is underway. You have been given much to work with. How serious are you? Are you committed to this path, and are you willing to do the inner work necessary? Are you willing to let go of ego control and discipline yourself to meditate daily?

It is easy to do this at the Mystery School where it is part of the curriculum but how about out in the world? Are you willing to act decisively on your inner guidance? Do you spend time providing yourself with a quiet space where you can contact your inner self?

As mentioned previously, there is plenty of time. All these things will come about for you. In which incarnation depends entirely on you. This book is written as a catalyst and stimulant, as you are experiencing. It has many, many levels.

There is an order to the cosmos, to the Divine Creator's plan, and the plan is executed through the stepping-down of energy. Humans have the ability to learn about and experience this energy and to use it consciously for their evolution. Spiritual evolution occurs when one integrates love.

At this point of human evolution there is a window of opportunity being provided for all. The window can also be called an awakening. There is an opportunity for each individual to awaken to the understanding that all are connected to The One. All carry the same energy. All are connected as One.

The path to this integration is different for everyone. Each individual chooses his or her own path, either consciously or unconsciously. At some time along the way, the individual becomes conscious of the many facets of self and this brings awareness. From this self-awareness comes choice.

How do I play my game of life now that I am aware? Do I choose to play according to my past patterning, or do I choose to play a loving and constructive game? How shall I co-create with The One as my partner?

You, as probationers have an exciting path to follow. Where will it lead? Why not choose to consciously and constructively utilise the energies of the great Rays, beaming through the zodiacal constellations to assist? Your path will unfold according to your soul plan. Your birth chart is its model.

It is through constant study of the cosmic energies on a daily basis, that you will learn to feel how they work within you. Intellectual knowledge can go so far, and then it needs to be applied to your life. You have all the tools available to you. How wisely do you use them?

These seven Great Rays, these beings of Light, play their part. Their love for you knows no bounds. By working constructively with the theme of each sign, feeling within you as you apply this energy to your life, destructively or constructively, and then consciously choosing to apply the positive energy, you will begin to access the higher vibration of the Rays.

Each Ray has a divine purpose leading to human evolution. What a great service these Lords render. Why not choose to work with them?"

*

A notation: D.K., in Alice. A. Bailey's Esoteric Astrology book, says that Libra transmits the potencies of the Pleiades, and Aquarius expresses the universal consciousness of the Great Bear. Gemini forms a point of entrance for cosmic energy from Sirius. These three, form the greatest over-lighting triangle of force within our Solar System. Through these great systems, the Seven Rays enter our planet.

As I conclude this chapter on the Seven Rays of Creation, I remember the Hopi Indian Prophesy I was asked to write about years ago. When visiting Sedona, sitting on a log relaxing in the Sun at the airport vortex, a highly active area of Earth vortices in Arizona USA, I was approached by a Hopi Spirit Being. I'd never been contacted by him before so the experience was new. He asked me to write about the Hopi prophesies and their relevance to our current experience. This prophesy is in addition to the others I wrote about in *The Magdalen Codes*. It follows:

"On August 17th 1987, 144,000 sun-dance enlightened teachers will awaken in their dream mind-bodies, and the various winged serpent wheels will begin to turn, to dance once again, and, when they do The Rainbow Lights will be seen in dreams all over the world, and those Rainbow Lights will help to awaken the rest of humanity".

As with all types of ancient wisdom teachings, this Hopi prophesy is symbolic and allegorical, written for those who have eyes to see and ears to hear. At the time of researching the prophesy, it didn't make sense because I didn't have the consciousness to comprehend it. Now I do.

Many spiritual teachers began to awaken in 1987, around the time of the Harmonic Convergence. At that time, people all around the globe gathered at sacred sites to do ceremony and meditate for global peace.

The "wheels" of the prophesy refer to the human chakra system. Each of the seven chakras in the human etheric body begin spinning as an individual's consciousness evolves into greater Light through self-awareness. The "winged serpent" refers to the kundalini energy, coiled at the base of the spine that begins to rise up through the chakras when an individual develops higher levels of consciousness by willingly accepting self as cause of revealed errant behaviours.

This spiritual evolutionary process enables embodiment of the Seven Colour Rays of Creation that can be seen emanating from the aura, and the hands of evolved spiritual healers, The Rainbow Warriors, and the Rainbow children.

In my *Magdalen Codes* book I mention some of the many annual journeys I took to the sacred sites of Peru and

Bolivia, in South America. One of the sites is called the Aramu-Meru Gateway. It's a portal to higher dimensions. Aramu means serpent.

Lord Meru was a Light Being from a higher dimension who made himself known to me through this dimensional Gateway when I took a group there to dance the sacred circle dance Paneurythmy. Through the love and unity created by the group dancing, we opened up formerly closed interdimensional knowledge portals.

During this Age of Aquarius, Ray Seven, a violet colour, is a Ray now permeating our planet and humanity upon her. Ray Two, the Ray of Love and Wisdom, is always beaming on humanity, stimulating us to merge with the vibrations of the unified body of Christ Consciousness.

On the 13th February 1988, six months following the Harmonic Convergence, at 12.30 am G.M.T. an exact conjunction of Uranus (Universal Light) and Saturn (Time Keeper) occurred in the final degree of Sagittarius, a fire sign, penetrating the magnetic fields of the Earth. It ignited a crystal transmitter cell beneath Uluru in Central Australia. A pulse of love energy flowed through the harmonised consciousness grids of the Earth.

Uluru is one of the key planetary vortices for the regeneration of world cycles, hence the beginning of a new Age must begin with a transmission of energy from this point. This fire energy ignited those souls ready and willing to do the inner work to awaken to higher levels of consciousness.

On December 17th 2020, another conjunction between these two planets occurred, in the final degree of Capricorn, an earth sign, anchoring the Christ love energy into the planet. Many gathered at Uluru for ceremonies, prayer and meditation. A brilliant blue/white light was seen in the sky at the exact time of the conjunction.

The Hopi Indians believe that they are descended from Pleiades and even before that from Lyra. The call Lyra the "Eye of God". They believe that spiritual love is the answer and that comets are a sign of prophesy.

They also say we can reactivate our 12 strands of DNA ourselves. Hopi's are the record keepers of the native north Americans Indians. They say we are entering the "Fifth Age" the Age of Illumination.

They also say that Sirius is the blue star. *'Kachina' dances in the plaza and removes his mask.* The mask is the mask of the subconscious. Isis is the feminine deity associated with Sirius. A return of the higher frequency energies of the goddess, the divine feminine archetype, is connected to the new consciousness. Blue = electricity. The energy of Sirius is blue (higher consciousness that acts like electricity)

Ref: Author, Teacher and shaman – Robert Ghost Wolf. August 1997.

*

PART TWO

The Great Bear Awakens

Bear enters the womb-cave to hibernate and introspect. Allowing his intuitive mind to reveal truth, bear enters the great void, the place of inner knowing. His ancestors and spiritual teachers sit in Council guiding him to answer his own questions. In India, this cave-womb symbolises the cave of Brahma. Other cultures refer to it as the Dream Lodge. When guidance is given and truth is known, the bear leaves his cave to begin anew.

Two Self-explanatory and Relative Transmissions follow:

And it came to pass that the Prime Creator decided to create aspects of Itself to further the Great Work. Planet Earth was the most appropriate place. The Prime Creator prepared, and when all was ready, the human experiment began.

Into the vast oceans were placed embryos of the species that would become human beings. These embryos were encased in cocoons that would provide a safe haven until the cosmic timing stirred them into action. Joined together by a pulsating living light cord, the embryos grew. The ocean floor provided the safety they needed.

At the cosmically appropriate time the cocoons began to open as each individual being desired emergence. These embryos, able to breath under water, made their way to the shores. Eventually they grew legs and were able to walk upon the earth. Nature supported them, and all their needs were met.

These first human beings loved and respected Nature for all it freely gave to them. Harmony reigned. The earth was filled with the fruits needed to sustain human life. Human life was treasured and revered.

Three areas were first populated. What is now Australia, Peru and Africa. The lands were different then. The oceans did not separate these three countries. They were one. The warm ocean had nurtured the first humans well.

This early species thrived. Other cosmic beings heard of the experiment and wanted to serve the Creators' Great Work.

They played their part in the human experiment. Some of the most ancient myths, stories and legends tell of these times.

Other experiments followed. The triangle of force prepared other cocoons with the precious occupants in the northern hemisphere. The human developmental process is still unfolding. Each human being was given the gift of free will.

And it came to pass that beings from other star systems heard of the Great Work and volunteered to help. The volunteers planned their explorations with their leaders until all was in readiness. Different areas of planet Earth were chosen according to the purpose of the planned programme.

These volunteers were able to adjust their vibratory patterns to suit their surroundings so the exploration succeeded. In the early stages of the experiment the genuine desire to assist the Divine Plan was the motivation behind its success. Humans evolved accordingly. Love was the force behind The Great Work.

The light prevailed.

Where there is light there is also dark. With impure hearts and with motivations to get and take, a shadow enveloped pockets of the earth. There have been many stories written about this phase of human development. Pockets of light balanced the shadows and overall harmony still existed. When this balance is broken, catastrophe occurs.

It is, and always has been. Earth is a plane of duality.

A trinary system is being prepared to overcome this situation. For too long Mother Earth has suffered from human abuse.

Humans are now awakening from their deep amnesia to remember their pristine origins and the paradise they inhabited so long ago.

The Creator gave the gift of free will. Enough people are choosing to use it to raise their levels of consciousness and to follow the call of their hearts. By feeling into this call, and acting upon it, The Great Work can be accomplished. Every human who acts upon the joy of their heart and soul is pre-programmed to play an important role in the resurrection of light process. A new Camelot is gestating.

From the distant past to the future, the human evolutionary experiment is being guided by the Galactic Federation, a host of celestial beings who are intent on carrying out The Great Work. Many humans are aware of portions of the Plan, and are contributing to it. We, in the higher realms value their contribution.

You, Ashtara, are the only Sirian ambassador on Earth, and as such have an increasingly responsible role. We will only guide you to your capacity. Never doubt we are with you always. Your work is to accelerate at a pace we know you can handle. Accept we have your best interests at heart.

You have explored the galaxy with eyes wide open and shared your findings as was asked of you. Now your work is more earth-bound. We ask you to trust our guidance. Resistance to the Plan will create problems. These will be easily overcome once acceptance is integrated.

We honour you for your continual service. We are always with you and love you dearly. We are your brothers and sisters from Sirius.

*

The Arcturians have the following to say:

Once upon a time when Earth was young, many beings from other star systems enjoyed their holidays here. It was a paradise, a haven, a respite from their normal lives. These beings, of semi-etheric form, appreciated and valued Gaia's beauty and bounty.

A Great Plan was created by which a species of life form would live permanently in this wonderland. It was decided to impregnate an already existing species with certain formulas that would enable this particular life-form to evolve and grow in accord with nature. This life-form became known as human.

Over aeons of time the species was impregnated with different codes that were dependent upon which visiting species worked the experiment. Humans today are a result of these genetic experiments, with the ability to genetically modify other species just as you were once genetically modified. The wheel has turned full circle. Those who implanted the genetic material can now see clearly the results of their work.

In the 21st Century, implants are commonplace. Some of these implants are used for research in an attempt to understand animal behaviour and instinct. As was once done to humans, humans are now doing to animal species.

The zodiacal system in use today is a system of duality. Prior to the tampering of this system there was a trinary system. It was this trinary system that created the paradise on Earth. The understanding of this trinary system is known to some people, and this knowledge will spread. It is time for humans to wake up from their amnesia to know the truth of their heritage from the stars.

You are a mixture of genetic material from other star systems, and you are an experiment. You are a genetic experiment and, as such, are mutating into distinct polarities. In your zone of duality there is always darkness and light. The battle between the dark and light is intensifying. The dark will do its best to maintain its position and keep on perpetuating amnesia. It does this through the media and misinformation.

The Light is doing its best to awaken humans to truth and it does this through love. The self-generation of love within human hearts and minds is a spiritual technology that is now being practised by millions.

As this wave of love sweeps across the planet, its light will become too bright for the dark to handle. The dark forces operate behind closed doors. The Light operates in nature where there are no doors. The dark desecrates nature. The light honours and respects nature.

Within the hearts of all humans is a divine spark. This spark becomes a strong flame when the decision is made to follow the Way of the Heart. Heart intelligence slowly begins to take supremacy over the mental aspect. Heart intelligence is connected to the Laws of Nature.

Throughout the Ages of human existence there have always been great spiritual teachers. All have said the same thing. The living light of love is the necessary ingredient to fuel a healthy and fulfilled life. It is so simple. The living light of love does not abuse, maim, destroy, hate or fight. The living light of love accepts, respects, and appreciates all of nature for its inherent teachings. The Way is easy to follow when people are awakened and aware of nature's wonders.

The early morning birds sing their song joyfully and their high vibration sends waves of love to the plants. They open their leaves and flowers to the new dawn with eagerness and joy. Where nature is encouraged to flourish the light of consciousness grows strongly in the soul. The opposite is true. Nature uplifts the human spirit. Will you allow it to uplift your spirit?

Preparatory Celestial Training

During regular meditation, knowing that the chakras are the only entrance into the human body of cosmic light, I clearly visualise opening and spinning them, from the base to the crown. I then visualise myself standing on an etheric platform above my crown chakra where I view space. I do not allow extraneous thoughts to enter my mind.

Instead, I focus on generating divine love in my heart, observing myself feeling into every tiny vibrational nuance in my body and mind while doing so. Calm, deep and regular breathing assist the process. I invoke Cosmic Light, and visualise, see and feel its entrance as a pillar of Light into my physical body via the spine and chakras.

Many years ago, during meditation practices, I was trained in mental telepathy, bi-location and remote viewing. The training was given by love and light filled Spiritual Masters, Archangels, the Elohim Creator Gods, and highly evolved extra-terrestrial beings, in particular the Group of Nine from Sirius. For astrologers reading this book, my birth horoscope Sun in Cancer in 8th house conjuncts Sirius within minutes of exactitude. Mercury closely conjuncts my Sun.

I became adept at space travel. It felt natural, as if a way of life I've always known. During the training I was taken, first by the Archangels, Uriel and Gabriel, and then by other cosmic transport means, to various star systems. My invisible guides asked me to record the experiences and write books about them.

During the time transiting Pluto conjuncted Antares at 10 degrees Sagittarius, I was trained, during meditation, by a group of light beings from Antares. I regard it as a great privilege to have been trained in this way. Some of their training experiences are written about in the first book of this Arcturus trilogy, *I Am an Experiment, and Extraordinary Spiritual Adventure.*

Others now follow, as simple unedited journal notes. In hindsight, I realise the experiences I had were given in preparation for my readiness to receive training in the Arcturus system of Astrology.

January 1st. 2007. New Year's resolution: To meditate more regularly.

During this New Year meditation, I visualised myself standing on what I call a Way Station about a metre above my crown chakra, where I completely blank my mind after asking for contact with my Higher Self. I waited, without expectation or thought, for quite some time, totally trusting. Two loving Archangels entered my etheric space, embracing me with their wings.

They took me into space, seemingly a long flight. I saw beautiful aurora lights and wave forms of pink and purple dancing and projecting outwards, inviting me in to their centre. As I entered, the coloured wave-forms merged into one single colour.

The one colour was all around me, permeating my whole being. I was asked to sit. I became aware of an unveiling taking place in front of me. It was as if fine, very tall curtains of gossamer light parted. I saw waves of pulsating light forms coming towards me from behind the opened space.

The light forms entered my body form in a line, one after the other. I felt their pulsations inside of me. The pulsations attuned to my heart beat and became waves of energy. I observed and felt. Then the curtain dropped and the pulsations ended.

I was asked to swivel 180 degrees to face myself. What I saw was of giant proportions. I couldn't see the top of me because I was so huge. After I had faced this huge self I was swivelled around again to my former position and advised I was on Venus, and that I had now merged with Venusian energy as I had done with the other planets. Jupiter and Uranus were to carry me on to greater heights.

I was then taken back to my body and it was as though I was gliding, much like an elegant swan. My Higher Selves are angelic beings who said they would introduce many opportunities. My task was to be discriminating about the choices I made. To do so, I needed to centre myself and to feel into the opportunities presented, and only act upon those that felt true to me.

January 2nd

I am again taken up into space by the Archangels, Gabriel and Uriel. I felt psychologically dense, so we stopped along the journey to enable the angels make the adjustments needed to my energy field. It wasn't long before I observed purple light all around me and a feeling of warmth within my heart.

Continuing on, we moved up into greater light, the light of the Sun. I did not feel external heat, but felt love within my heart. There I was introduced to RA, who, I was told, is one of my spiritual guides. He appeared enormous and radiant. Dressed in Egyptian style clothing he radiated love, power and brilliant light.

I asked about Isis, and was advised I was not yet ready to meet her. RA advised he had been with me for some time, however, today was a formal meeting. He is the same RA as my former self, Barbara, experienced on her visits to Sirius.

I then realised I had allowed myself to slip back down the spiritual ladder into arrogance; which is a cover-up of rejection, a lesson I still needed to learn and alchemise. Rejection hurts. Vulnerability can be scary. I am determined to learn my lesson. Planetary cycles repeat. I need to learn from them.

January 3rd:

At my Way Station, the two Archangels wrapped their wings around me and I felt nurtured, supported and inspired. They showed me how my spine was now a "feathered serpent" and that my wings would emerge from it. They asked me to practice opening them.

I did what they asked, and the white wings that emerged were enormous. So light and soft. I encircled myself with my wings, feeling so nurtured. The Archangels told me that I was now ready to fly like an angel, suggesting I visualise myself as a giant white butterfly in order to do so. They said there would not be a flight with them today, and that today is a day to remember. It's a day when I accessed my own wings and felt their presence within me.

They asked if I had any questions, and I asked some personal ones. Their answers were illuminating.

January 4th.

When I reach my Way Station, I saw and felt my wings unfold naturally, without any thought on my part. They are so huge. The two Archangels appeared and took me flying, advising there would come a time when I would be able to do it on my own, and would no longer need them. Right now, I was in training.

As we flew, I was asked to report on my experiences. I answered saying that I felt no form, and had a myriad of tiny single stars glistening without visible connection one to the other, yet somehow all connected. We were like a wave-form, working together in synchronicity just as a flock of birds fly together.

Birds have no apparent visible connection, yet are connected by some invisible built-in radar thread. They fly as one group consciousness and act in unison instinctively when it is time to turn or land. I felt as if the space inside of me contained a myriad of tiny lights, flying in unison as a wave. I could not feel my physical body.

The Archangels appeared to be well pleased with my observations and, as we returned to my Way Station they asked if I had any questions. Again, I had personal questions and their answers were illuminating.

January 5th.

I was taken up to experience the light of the Sun. It felt so warm in my heart yet had no external heat. The light was incredibly bright, much like the light that comes from someone welding steel. It was blue/white light and was so bright I had to close my eyes. However, I could still see the bright light even with my eyes shut. I was guided to return down to my Way Station.

January 7th.

My two trainers, the Archangels Gabriel and Uriel, took me up into space again. Just before we arrived at our destination, a special fountain, I felt dense energy begin to stir in my solar plexus chakra. When this occurs, it always feels nauseous, as if shadow psychological energy in preparing to reveal itself. Some energy is also stirring in my field, connected to my throat. I can feel the dense energy from my solar plexus move up to my throat chakra.

I am asked to step into the clear water of the fountain, contained within a pond, then asked to step under the fountain. I am then baptised into *The Way*, and anointed into *The Way, The Truth and The Light*. I hear the words "*May the love that binds you always bring forth truth and the light of wisdom*". As I return to the Way Station I hear the words "*Rest today, for tomorrow the work begins*".

January 8th:

I am at my Way Station and asked to fly by myself. With wings open I lift up from the station platform with relative ease, realising I'll need a lot more practice before I can fly alone and unaided. My wings are so huge although they feel as light as a tiny feather. The Archangels ask me to ponder on the question "Who Am I?", and to define it.

I attempted to fly and used my focused mind to lift up. I was aware of Gabriel and Uriel beside me and experienced them as supportive "training wheels". I managed to lift myself up quite high, but faltered as soon as I looked down. I caught a thermal current and flew/circled with it. It felt strange, but somehow familiar.

My wings are so huge in comparison to my body and spine. I realise I am in training, but for what?

January 10th.

My Higher Self Archangels took me on a journey into space, to a place of incredible light. At first, I had difficulty in seeing what was contained within the light. It felt so comforting and familiar. I looked more closely to make out an assembly of seemingly infinite number of angelic beings standing in concentric circles all around me.

Their energy was loving, welcoming, accepting and calm. I felt I had come home. There was no other place to go to. I had come home to my angelic roots. I was home, to myself. My Higher Selves asked me to ingest, digest, accept and assimilate this totally natural experience.

January 11th

During meditation today, I experienced another practice with flying. I was shown how to lift off by generating the feeling of love within my heart. When I was able to self-generate this love frequency I could feel myself becoming much lighter, and lifted off successfully.

January 12th.

Immediately I was taken up from my Way Station by the two archangels to experience space. This was a totally different experience to any I'd had before. I felt I was invisible, that I was pure light and spirit, without form.

Before I took off, I felt my physical form and my body at the Way Station. I felt my feathered spine and my wings expand, but then something happened as I ascended. I saw the space around me as a black void; the black was far distant and there were particles of fine matter, seemingly moving on gossamer threads, around and through me.

I did not see any stars or luminaries. Yet I could see, sense and feel. It felt real. The Archangels asked me how I felt. I replied, saying I felt totally safe, secure, at home and curious. I was told I was a quick learner and that I experience as Spirit experiences.

January 13th

The Archangels showed me how we are one but we are many. We, the trinity, merged into one, then opened up and each one of us created three aspects of self, one for each point of a triangle. Then we merged back into one again. This is a repeat lesson to one Barbara, my former self, was taught on Antares. Note: I need to understand and integrate this!

January 14th

The Archangels asked me to created aspects of myself to form a circle. I did so. I could see a circle of angelic beings with wings touching. I was asked to count. There were twelve of us. I was asked "What was inside the circle?" I replied "Nothing, but everything. Great Spirit fills the space. Great Spirit permeates everything and is everything. It appears as a no-thing but is in fact everything, All That Is".

January 15th

RA came today, as a masculine aspect of my Higher Self. He radiated his light brightly, saying I was at a crossroads and could go either way. I could go back to the dark or onwards towards the light, i.e., the material or the spiritual worlds. He said that Barbara had created a bridge and had been able to manoeuvre her way carefully across the bridge. Will I do the same? Prior to this message, he said how he, as a spokesperson for his soul group, accepted and appreciated my love and dedication to the light, and thanked me for it.

Note: I need to quickly develop some very clear distinctions because I do not want to slip back into the dross and density of the purely material world ever again.

January 16th

I feel very clear and centred today. At my Way Station I was guided to place an etheric merkaba vehicle around me. I saw myself suspended above the Station as if by a thread. I was guided to spin the merkaba, with myself as a seed of light in the centre.

The merkaba vehicle was hesitant. It only wanted to spin clockwise and then erratically. Whatever it did ended up successfully because I was then guided to dissolve it, and then open my huge wings. This I did easily. I was then asked to fly. This also I did, unaided. I hovered in space, not knowing where to fly to. Then the idea came to fly to Venus. This was so easy, so very easy.

I flew as if I belonged in space. Venus appeared. Her base colour of mountains, valleys, lakes and rivers was obscured and veiled by a mist of pink and violet waves. It was beautiful, and looked elegant. I flew over it, feeling awe and wonder. I didn't want to stay long because I somehow knew I could return to explore at some other time.

When I returned to my Way Station I descended into my body, feeling so filled with love, grace and peace. I toned my special benediction - a series of vowel sounds that spontaneously erupts within me when my vibration has increased to a high frequency. The toning is a resonating connecting devise to my soul. It was a beautiful experience. I said a prayer of gratitude.

February 2nd

At my Way Station. I was taken up high into space and asked to spread my wings, so I did, and flew to a beautiful lake in the mountains. This was Lake Titicaca in the high Andes Mountains of Peru and Bolivia. I landed on the Island of the Moon. I was not alone. I was Lady Meru, and Lord Meru was with me. I was shown how we were Archangels, and that the lineage of Mary Magdalen was from the Archangels. I wondered how the mermaids from Sirius and the Archangels were connected?

Later in the evening: At the Way station I was asked to sit on the platform and t look down into the space below. I saw nothing but empty space. I wasn't afraid, seemingly sitting on a ledge. The space felt familiar and comforting.

Six Archangels gathered around me and asked me to watch as they were about to show me something. They were going to unveil something I needed to see. I wasn't clear about the next part, so I guess I must have gone into fear.

They told me they had the task of totally healing my Leo/ Aquarius separation from love psychology if I was willing to allow them to do so. I was amazed, and so very grateful. They said they were not in a position to heal all at once but I would feel the difference and know. They said they'd been instructed to do the same procedure on each forthcoming Full Moon and all I had to do was to make myself available. I made notes in my diary to do so.

*

Arcturus Training

From February 7th 2007 I was trained by the Arcturians to learn a new astrology system, the answer to my continual question:

"*What cosmic system is available that continues to promote human consciousness evolution after we have worked through our Solar System's planetary challenges and themes depicted in our birth charts?*"

The Arcturians asked me to record and to share my journal notes relative to their training. Each guided journey to Arcturus, an etheric Mystery School operating in a higher dimension than that of planet Earth, took the form of a visual, telepathic and experiential lesson.

Following are my journal entries and my experience of the system I was taught during these meditations. Many relevant Arcturian transmissions are included.

Prepare yourself for take-off.

*

From my Way Station, about a metre above my crown chakra, I was guided to fly to Arcturus. Arcturus is part of the Great Bear constellation and is connected to King Arthur and the Knights of the Round Table. On arrival at Arcturus I was greeted by a Light Being who invited me into a boardroom where I was immediately warmly welcomed, as if expected. Observing a round boardroom table in the centre of the room I was advised "This is where the Masters work".

I observed a zodiac engraved into the top of the round table with a Sun at its centre. The zodiacal images were pictographs. A group of earthly students were seated with me around the table. I intuited they were mostly from different areas of Australia.

"This is the Wheel of Life", we were told. I was asked to look above to the ceiling where I observed another zodiac with the three bright stars of Sirius in the centre. This zodiac was beaming light rays through triangular pyramidal forms to our zodiac of twelve seemingly engraved on the boardroom table. This, we were told, was a system of cosmic management. After ascertaining I had understood what was being shown I was dismissed from the class and made my way back to my Way Station.

That was apparently enough for me to observe and absorb for that day.

*

A few days later I was guided to inter-dimensionally fly to Arcturus again. The class was in session with students seated around the boardroom table. I looked carefully at the zodiac on the table. It was mirror reversed.

We were taught that, once the karmic wheel slows and stops spinning, and internal balance is attained between our inner masculine and feminine, the rotational movement of the soul's journey changes. From then on, our soul progresses clockwise from Libra, and the influencing zodiac is reversed.

The movement is from Libra and return, via Virgo. Libra is the balancing point. Balance comes from within when our internal divine feminine and masculine energies become synchronised, and when personal responsibility is taken for all we create. Our soul is responsible for the timing of this evolutionary directional change.

We, as a collective human consciousness, have been in the darkness of ignorance since the last Age of Aquarius.

When I viewed the ceiling, complete with the stellar system, I noticed the light beams from this greater zodiac were steady and constant. There were three main ones. My intuitive understanding was that the Atlanteans created the current modern astrological model in order to manipulate humanity's fall into density and ignorance. It was done as a method to gain power and control. It is now time to change the system back to the original Lemurian blueprint.

At the conclusion of this important and revealing lesson I travel back to my Way Station and re-enter my body. My mind immediately goes to the story of Isis and Osiris. Isis was a High Priestess, who, at the end of her earthly life, returned to Sirius.

My question was, "Who brought the Atlantean knowledge to Egypt?" "Thoth, the monkey God", was my answer.

I recall a past life in Atlantis when I was an intuitive young, fourteen-year old boy who saw and understood the power plays taking place in high echelons of scientific power. My father was a senior scientist working with the giant generator crystal. I knew his superiors were manipulating the crystal power for their own ends, and advised my father.

He didn't believe me and mentioned my understanding to the authorities. They had me imprisoned in a hanging cage, situated over and between two narrow chasms. Sometime later, he became alerted to the truth of my knowing and realised the destruction of Atlantis was imminent. He made the decision to negotiate for my life at the expense of his, blackmailing his superiors into providing a safe passage for me to Egypt with them, should their crystal experiment prove disastrous.

Thoth was one of the members of that space transport. After we arrived at our Egyptian destination, Thoth manipulated his way into the esoteric temple circles where Isis and Osiris were teaching pure Sirian wisdom.

Osiris was killed and chopped into pieces by his brother, and Thoth helped to restore Osiris to life through his magic. Arachne, Spider Woman, saw what was happening, and swallowed Osiris' phallus. Thoth was thwarted in his mission. He could only hold the focus of darkness for approximately twelve thousand years.

The collective feminine, the serpent power, would uncover the secret through intense focus and penetration into remembrance. The feminine would reveal the truth when the time was right.

Gaia's role of being a galactic garbage dump for human negativity was coming to an end.

*

On the next visit there were twelve students standing around the zodiacal round table. The teacher made thirteen, a number attributed to birth, death and rebirth. The teacher told us that the reversed zodiac on the table was the correct one. It had been changed in Atlantean times as a power and control method to keep humanity in darkness and ignorance of truth.

Immediately I felt myself lose focus and was asked to look up. The beams of light seemed to emanate from a gently moving prism. The brilliant light appeared to move, not as a lighthouse beam but as small slight movements. This observation steadied me. I regained my focus through spiritual will.

The teacher said we were pattern breakers. We were to research, practice and experience so we would know this information was true, and then share it. When the time was right we would be called back to Arcturus to learn more.

I experienced a flash of illumination. A vision entered my mind of when I first visited Pisaq, a sacred site in the Andes mountains of Peru. When there, I recalled a past life of being an astrologer. My task at that time was to manage the cosmic energy so it was continually in a state of peace and calm. This was needed so the young trainee priestesses could experience their temple training with ease. I knew how to do this because I had learned to master the influencing cosmic forces. My tools were twelve special stones, not of the earth.

*

The next day, I was asked by the two Archangels to follow them as they flew to Arcturus. When at the board room I was asked to look up. There I saw a triangle of energies pouring down to Earth. I saw this as an equilateral triangle of three streams of light. On looking more carefully, I could see that at each point of the triangle were three streams of light radiating a total of nine rays of influence.

Then I saw a giant hand reach into the centre of this triangle and grab the sticky substance and pull it out of the centre of the triangle. This energy was held in the giant hand.

The triangle changed its location in space. It became distorted. The triangle swivelled, and the three points were then on a flat plane, held aloft by the hand. This image was very clear.

I was then asked to look at the zodiac on the table. It now had a white sheet placed over it. This sheet totally covered what was underneath. Over the sheet was placed the zodiac we know today.

At some point in time, the original blueprint for humanity's evolution of consciousness programming was covered over. There was a cover-up.

About ten days later, while at my meditation Way Station, my guides came and took me to Sedna, in our solar system, for healing. They referred to her as a star because she generates her own light from within, something we humans are learning to do. I experienced fine frequencies of mauve and pink and felt balanced and calm.

A few days later, I experienced a healing from Chiron. He asked me to write again with my non-dominant hand because it connects to my heart.

Two or three days later I flew again to Arcturus. My stomach felt nauseous. Energy was stirring. Was I afraid to uncover truth? Were more memories stirring?

I sat in my place around the table, observing a lesson taking place. Again, there were twelve students around the table. I firstly checked to see if the light above was beaming down. It was, but appeared different from before.

This time the beams appeared to emanate from a triangular or fan shaped formation and beamed down into a single focused light, seemingly opposite to what I had seen previously. I was curious about this phenomenon.

Our teacher revised the lessons saying that the reversal of the wheel was allowed to occur as part of the divine plan. There was an opportunity approximately 12,000 years ago, during the Age of Leo, to correct the imbalance of human negativity, but it wasn't successful. All it will take this time around to correct the imbalance is a handful of people willing to spread the word. Who was willing to do this? I put up my hand, as did only a few others. The teacher said this was enough, and it would be done. It was time.

Then he explained the reversed phenomenon. Saying, "We on Arcturus are now the foci for beaming light to others. There was a rotational system of energy and Arcturus received the light emanating from other sources, three main ones, and then transmitted the radiation to other star systems. It is an energetic system".

Some days later, I returned to the Boardroom table. We were told that the zodiac we have been using since Atlantean times was created by Thoth and is a distortion. It is a cover-up. He used his knowledge to deliberately distort and change the correct system. Tomorrow we are all to assemble again and will be shown the correct system. Some already know this, and are working with it.

A little later in the day, I had a message from one of my Pleiadian guides. She said, "We in the Pleiades have accepted our karmic burden. We choose to correct the harm that has been done. One way we can do this is through the illumined minds of astrologers. You are about to reveal a deeply hidden truth. A secret that has resulted in the near destruction of our experiment. We will not allow this destruction to take place".

"We are prepared and willing to assist you in spreading the truth, in different languages. You are our creation. You are our babies, and we abandoned you. The deep feeling of rejection you experience within your souls stems from our abandonment of you. It is time to heal this. We are committed to this path. For those who have their eyes, ear and hearts open, we ask your forgiveness".

I digested these words and then had the illuminating thought: "What if my 'walk-in' experience was purposefully timed in order to experience deep rejection so I could feel it, work with it, and then connect the dots on a cosmic level?"

And this is so, said my guide.

21st April:

Pausing at my Way Station, I prepared myself to fly. I opened my wings, feeling them emerge as my consciousness focused on them doing so. They were white and incredibly light. I saw myself as a very thin body with gigantic soft wings. I generated my fuel, that of Divine love, and took off into the unknown. I found myself almost immediately at the Arcturus boardroom. I sat in my appointed seat around the table with the other students.

We were thanked for our attendance and asked to wait quietly. A flustered Frenchman soon occupied the empty seat beside me, and the session began.

Looking up, I observed a different light beaming from above the table. The spokesperson advised he was about to uncover and reveal the true system, one that had been covered over since Atlantean times. The cover-up was extremely clever, and through it, humans have been manipulated and controlled. It was time for truth to be revealed.

As the sheet was rolled back, a brilliant picture emerged. In the centre was a living bright pulsating heart. The entire model appeared to breathe. The light around the heart was bright but not blinding.

I looked closely and noticed small cosmic bodies orbiting around the centrally positioned pulsating heart. There were nine satellites, or cosmic bodies, around the central heart hub of the wheel, seemingly attached by spokes. The entire system was alive and breathing. I looked up. The light above felt and looked like soft and luminous moon light, or maybe it was starlight. It felt feminine.

This was the end of the lesson. We were asked to return tomorrow.

I again journeyed to Arcturus from my Way Station. Sitting at the round table with the other participants I noticed, and felt, how powerful was the pulsating heart in the centre of the table. The energy coming from it was pure love. I felt it activating and filling my heart. We were told how humans were created in the image and likeness of God and the image in front of us represented the Creative Force. We humans were created as pure pulsating love.

Sirius is a representative of the Creative Force that governs, through love, the sacred geometry in our part of the Universe. Universal love holds the sacred geometry in form. I could easily see how the energy of love was conducted from the central hub through the spokes of the wheel to the other satellites, or sub-stations.

The satellites were fuelled by love and the intelligence of the heart. We were then shown how a sheet was covered over the system to prevent the intelligence of the heart from being distributed further. Another ingredient was added. This was the substance of logic.

Class disbanded due to static. From my experience and understanding, frequency static occurs when human thought goes into over-drive and the resultant emotions – such as fear or confusion - run high.

*

To have a break from this information, and to enable integration, following is a related Arcturus Transmission. But first, some wise words from physicist, J.J. Hurtak in his monumental book *The Book of Knowledge – The Keys of Enoch.* He writes that Arcturus is a major programming centre of the Galactic Council serving the Father on this side of the galaxy, which in under the direction of the Council of Nine, the governing body of our local universe.

*

From time immemorial the forces of Light are always working to overtake the forces of darkness. At different times in the human evolutionary experiment, interference by the forces of Light was needed for the survival of the human species. Interference by the Light in the natural evolutionary process is very rare.

A wise mother guides her child and allows it to make mistakes and learn from them. She interferes only when the child is in grave danger. This is the process of the forces of Light. A child will usually test all the parental boundaries. So too does humankind test the natural Laws of the Universe.

Man has walked on the Moon but continues to pollute the atmosphere. This pollution is approaching dangerous levels and it may not be too long before it threatens the survival of humankind. This is being monitored very carefully.

The force and power of one man destroyed millions of people during the second world war and contaminated the minds of many more millions. That contamination is still alive and well.

Yet the forces of Light never give up. Their numbers are increasing as more and more humans wake up to the truth of their heritage.

One of the greatest perpetrators of darkness occurred in Atlantean times – in pre-history, and came through the brilliant mind of one man.

The Sirian system of knowledge radiation had operated successfully for aeons of time. The Lemurian civilisation thrived on it. At that stage of human evolution, the worship and honouring of nature was well practiced and natural laws adhered to. Light was the predominant force.

When it was time for that civilisation to move on, many wise teachers travelled to other lands carrying the knowledge of the Laws of the Universe. The Sirian system of cosmic radiation was the basis of their knowledge. Some of these travellers chose to live in Atlantis.

The Atlantean civilisation was working with technology and began to disregard natural Law. Some of the brilliant minds realised they could control nature through the abuse of natural resources. That technology developed to even greater heights than it is today.

However, like today, greed and power reigned supreme within the collective consciousness. On the surface, the cities appeared unpolluted as they wisely chose to use natural means for generating their power. Their knowledge, and use of crystal power was far greater than is technology today. Crystals were plentiful in that land.

Prior to the demise of Atlantis, some people travelled to Egypt taking with them the knowledge gained through experimentation. In ancient Egypt, the Sirian system of cosmic radiance was appreciated, understood and revered. Rulers worked with the system and the civilisation thrived. The visitors experimented with this system and applied their Atlantean technology to test the system for their own use.

Geometry played a major role in the cosmic cover-up.

*

The Intelligence of the Heart

Now, let's return to class.

At the next lesson we were shown how the pulsating heart of Sirius B guided the heart of Earth into synchronicity. Heartbeats aligned. We were asked to learn about each satellite, and told that the Sirians and Pleiadians were the main species responsible for implants into the human mind. Interference to the natural evolutionary development took place.

Both of these extra-terrestrial races want to move on, and to do so they must first heal their interference karma. The Universal Law of Attraction applies in all instances. As with a human psychological wound, the way to heal is to access cause – reveal the truth contained within the creation of the wound, accept, and take responsibility for it; embrace it with love - and make the appropriate amends.

Sometime during this lesson, we were asked to look up. I could not see the source of light. The light beaming down on to the ten-heart system was as moonlight, soft and luminous, and enveloped the entire sky space. It felt comfortable, even comforting, feminine and nurturing – Goddess light.

*

In the following lesson we were asked to review the chapter in the bible about Adam and Eve – to read it carefully in order to extract the symbolism. I intuited that the apple

Eve picked is the knowledge fruit of life. And the serpent is the serpent of wisdom. The wisdom of the etheric kundalini force that lies coiled and dormant within the human base chakra. Usually, the invisible kundalini serpent is activated by accessing a soul memory leading intuitively to a major life-changing self-realisation.

The self-realisation enables personal responsibility to be taken for causing a challenging psychological issue in the current life. As we develop higher consciousness through self-realisations, this coiled energy naturally moves up the spine, like a serpent, until finally, in a self-realised individual, it burst forth from the crown chakra. The following transmission connects to the teaching.

Where did our innocence go? What is the significance of the story of the Garden of Eden? According to the bible, Adam was the first man, created in the likeness and image of God. He was created as an experiment in order that God should know Itself. And then, Eve was created by taking a rib of Adam and adding earth to it.

We have all witnessed the innocence of a child. The delight and wonder on a child's face, as he or she first encounters the beauty of nature. This can come about by viewing the night sky, a bug crawling or a leaf or a flower opening from a bud. The child wants to know why and how.

Why did God create man in his/her own likeness? How did this take place?

Prime Creator, God/Goddess, Great Sprit, or whatever other name is attributed to the greatest energy source behind all of creation, allows all expressions of this life force energy to exist. Where this is light, there is always darkness and where there is love, there is also hate. It is.

All is allowed within the Creative process. Humans can learn to create consciously. Most create their lives unconsciously, so live life as ignorant zombies. Life force energy is all around us and courses through us, waiting to be utilised however we choose.

There are natural Laws present within this free substance of life force energy and these Laws apply as a constant. In order to understand, and consciously work with life force energy these simple Laws need to be understood. The innocent and curious child would ask "Why are these Laws in place?" And the answer could be "To protect humans from total annihilation".

A child needs boundaries in order to feel secure within his or her life's playground. Humans also need boundaries. The natural Laws of the Universe are always in operation, yet so few understand them. Most are completely unaware of their existence. Most humans are not interested in learning about the Source of energy and the spiritual laws that govern creation. Where has childlike innocence gone?

How did God create man? What was the substance that was used in the experiment? Why did God create man? Did males come first or was this a cover-up perpetuated by a male-dominated society?

The spirit of Prime Creator permeates all of matter. Spirit is refined energy. Within this refined energy are minute particles.

These particles coagulate together to form mass. A great mind created humans, connecting specific particles into form.

In the beginning was the void – nothingness and stillness. And then, a tiny spark of light, of consciousness, permeated the mass of nothingness. The spark created ripples in the giant pond of nothingness. Disturbance, in the form of small ripples resulted. The spark that penetrated into the sea of nothingness was as a seed planted into the mind of God. Nothing could ever be the same again.

The seed has an opportunity to grow however it needs nourishment and water. Water is a carrier of consciousness. Consciousness is the substance of Spirit and contains knowledge of all that is. And, all that ever has been and will be.

This spark of consciousness can be likened to a seed planted into the great mind of God. An idea was seeded. Where did it come from? Was it planted by an even greater God? What is the beginning of creation? How will creation end? Will creation ever end?

Being a practical, earth-bound human, having had regular life-changing multi-dimensional experiences, I continually seek to make sense of my reality. I enjoy connecting the dots in my mind. When I do, something alters inside of me. Something I read, whether scientific, mystical, astrological, psychological or spiritual, from which I receive an instant resonance to the material, results in my entire body vibrating to the truth revealed. All my senses activate in resonance, and my soul remembers. The question is: How can I put words around my revealed knowing and felt sense?

The Zodiacal Elements

Maybe twenty years ago, when I first saw an illustration of the human energy field containing the three elements of fire, air and water as energy bodies around a human, I resonated with a truth. From then on, Lesson One in all my astrology courses demonstrated this model.

I used a green coloured stick figure of a human in the centre of a flip chart, symbolising the earth element. Surrounding the earth body in an oval shape I draw a blue line symbolising the element of water, representing the emotional body. Surrounded by the emotional body, symbolising the element of air and depicted as yellow, I draw the mental body; and the third outer body is that of Spirit, representing the element of fire, coloured as red.

My teaching is based on the understanding that, until we activate the fire of Spirit in our lives, and learn to consciously develop mental and emotional mastery, (emotions being caused by thought), our human body, the earth element, cannot function to its full capacity. All life begins with the etheric spiritual outer body. We are a spiritual being with a soul having an earthly experience. Our soul and spirit reside within our body.

Developing an understanding of our human elemental nature is an important part of learning, and integrating astrology.

To the ancient traditional alchemists, who could be considered the first scientists, (the second being astrologers), Fire, Air and Water were known to represent the extremes of polarity and balance. Fire was regarded as a masculine element, Water as a feminine element, and Air, the balance in between, was regarded as the child.

This trilogy connects to the Christian tradition of Father/Son/Holy Ghost. To the South American Andean and other indigenous cultures, the fire of Spirit is represented by Father Sky; The element of Air – by the wind, and Water - the oceans. Mother Earth is the recipient of these three elements. She is feminine, and receptive.

These elements connect to levels of consciousness. In the Andean esoteric tradition, there are three stone steps at many of the sacred sites, each depicting a level of human consciousness. The bottom step, symbolised by a serpent, represents the subconscious (water), the middle step, symbolised by a puma, represents the conscious mind, (air) and the top step symbolised by a condor, representing the super-or Christ/God conscious mind (fire). The foundation of 20th-century physics is the proton, electron, and neutron. Where am I going with this, you may well ask?

An equilateral triangle is a geometric symbol that can represent the three elements.

Arcturus Astrology is based on a system of triangles.

*

The Power of Thought

The Arcturians have the following to say:

In the beginning, the Lord God created heaven and earth and the animals that occupied the earth. Where did these animals come from, and how were they created? The power of thought creates energetic substance. Thought is energy. Energy follows thought. Manifestation occurs through the power of energy of thought. The more will-power behind the thought, the easier is the manifestation of thought into dense matter.

Is man alone in the Universe? Are there other life forms that also have thought? Do animals have thought? Why does an animal know what to do and how to do it in order to best care for itself? Is the action pure instinct, or is there thought behind the action? Domesticated animals seem to know their owner's needs, as if they can read their minds. Some animals seem to have a mind of their own.

The power of thought has resulted in genetic manipulation, a growing science. If there were other life forms within our universe, and it seems arrogant to believe humans are the only one of a type within this vastness of space, then it is certainly possible that these other life forms developed the knowledge to genetically manipulate humans.

Apes are akin to humans. For one advanced in the knowledge and understanding of genetic engineering, it would be simple to add a small percentage of genetic material to the fine substance of the ape's mind and thus create a human being. A being with the power of thought.

Why would a being of advanced consciousness choose to do such a thing? Why would they be interested?

Why do humans practice genetic engineering and cloning? Curiosity, greed and need form a strong motivational force. Curiosity, once fulfilled, can very quickly become greed.

Should there be other human-like species existing in other dimensions of time and space then they could also be curious, in need and greedy. They could, for example, need a particular resource that is either non-existent or is becoming depleted on their own planet. And so, they traverse space searching for this resource. They discover this resource, let's say it is gold, within the bowels of planet Earth. The question then becomes, "How can we extract it and transfer it to our planet?"

Clones are created through the process of genetic engineering and these clones extract enough gold for the owner's current needs. When the job is done, the clones are left to fend for themselves. They have been given the spark of mind by the 'gods' but have not be taught how to use it. This is their responsibility. And so, humans developed from this cloning experiment. They were created for a purpose, based on need.

I question and reflect. This may have been one way in which human life developed to what we are today. But has every human developed this way?

Earth is a large planet with many different countries and land masses. Over aeons of time the continents have changed. Earth has been here for millions of years. Different civilisations have lived on her. Some parts of the planet are much younger than others.

It is highly possible that different species of humans lived in different continents developing according to the specific resources of that land. It is also possible that species from other planets or star systems chose to live and play here.

*

The Great Cover-Up

During the next Arcturus class, we were told that Thoth, the one who created the current cover-up of the astrological system, had a brilliant mind. He enlisted the assistance of space beings to aid him in the process. The Pleiadians and Sirians worked with Thoth in the cover-up. Both species now want to move on in their evolutionary journey so must right this wrong.

As with a human wound, the way to heal is to access cause, and reveal the truth contained within the creation of the wound. This truth needs to be accepted, owned, embraced and loved, and then the appropriate amends need to take place. Thoth is often depicted as a baboon. Woven into the many tapestries depicting the Tibetan creation story I examined when visiting that country in 2006 were baboons.

Thoth is attributed to introducing writing, law and medicine to the ancient Egyptians. Mystical tradition suggests that Thoth also divulged the alchemical secrets of nature. His wisdom is said to have been recorded in forty-two books, later incorporated into separate texts, known as the *Emerald Tablets*. The knowledge from Egypt was assimilated by the Greeks, and Thoth became known by the Greeks as *Hermes-Trismegistus* (thrice born).

We, the earthly group sitting around the Arcturus round boardroom table, were told:

This is our action now,

Once truth is revealed, humanity is free to make appropriate choices and we are free to move on. It is time for humanity to evolve without the aid of their brothers and sisters from other systems.

The Age of Aquarius is the Age of Separation. It is also the Age of higher consciousness. The issue of separation must be seen in its true light before higher consciousness, based on divine love, can rule.

The human evolutionary process has reached a peak. The flow of knowledge will not stop. This knowledge wave will reach all areas of the planet. Wars cannot stop the process. Soon this will be realised by those who perpetuate the dark. The dark will continue to play their games but the trinary system of cosmic management will prevail. Duality, created by the cover-up, will cease to exist.

This is the plan for human evolution.

*

Why the Cover-Up?

On another day, I again travelled to Arcturus realising that the pulsating heart of Sirius and the nine spokes of the stellar wheel connect to different parts of the human body. I focused on generating love to keep me connected to the God/Goddess force. I feel plugged in while doing so. Just as any electrical devise needs to connect to a power source to be fully activated, so too does a human being. Source energy is our internal electricity generator, experienced as Divine love.

We were advised that my intuitive understanding of the model revealed on the table was correct and that each of us was to internally research to ascertain the organs each spoke connected to. We were told this information would not be found on the internet. It is how humans are created in the image of God and that Leonardo de Vinci had almost arrived at the truth.

On a Pisces Full Moon and total lunar eclipse, I was again called to the same boardroom table and noticed there were other shadowy figures already present. I sensed there were five of us. Their shapes and identities were unclear.

I was asked to feel into my heart and to see, sense and feel into which organ of my body the greatest flow of heart energy was going. I could feel it travelling to my solar plexus chakra. Within this chakra is another organ of perception – one far more accurate than the brain in the head. I intuited it is the belly, or digestive system. When a human being feels fear, the fear energy goes to the belly and it knots up.

I was then invited to leave the board room, and prepared to do so. I could see the other students intensely involved in their personal mode of learning. The French man next to me was writing. I understood each of them were being given information according to their particular way of receiving and integrating it. My guides advised they were proud of me.

*

The Arcturians have more to say:

How did the cover-up take place? The why is easy to ascertain. Power, control and greed are the ingredients for this mix. The desire to manipulate cosmic energy to prove one's level of magic to oneself can overtake any disciple on the return to the One. It is a free will choice, and your planet is a planet where free will is given.

The fall is great, yet so many souls choose this experience. It is a sure passage back to the Light. However, it can take countless lifetimes of darkness before the spark of light reignites.

A brilliant mind engineered a great cosmic cover-up. As with all creation, a desire arises within the mind. A desire to know and to experience.

In the beginning, the One desired to know itself and created images of man in order to do so. Focused intent and imagination were used in the creation process. From a soul and mind infused with thought and desire, the formation of matter occurs. It is the process of manifestation.

The brilliant Atlantean mind who desired to create the cosmic cover-up formulated his desire for power and control and imagined it to be so. He thought of all the details and, through his strong will, made manifest his envisioning. He became so consumed by his desires that he forgot the Universal Laws. His mind became clogged with density and lower vibratory thought forms.

All he had to do to play the game successfully was to align his body, mind and spirit to a one-pointed focus of purpose, and imagined it to be so. He abused his power and has suffered much as a result. The Law of Attraction is a given. It works in all situations.

The Universe is a system of energy. The One Supreme Creator, the energy of original sound, interpenetrates all creation – without exception - and accepts and allows creation. It is when the forces of darkness become stronger than the forces of light that intervention occurs.

There have been many past incidents of Star Wars. Planets have been blasted into oblivion and their soul fragments scattered throughout space.

On planet Earth civilisations have come and gone. Many civilisations have self-destructed, through abuse of power. As a result of this abuse, those experiencing the destruction fall into the murky mire of density. There are countless lives lived in an unconscious state, unaware of the creation game. They create scenarios of hardship and lack, depravation and poverty.

When the greatest depth of despair is reached, the human personality has nowhere else to turn. A desperate prayer arises from within the soul. The spark, long subdued, ignites. The long journey back to the light begins.

Desire, aligned with an idea, is the force behind all creation. An idea forms within the stillness of the human mind. This idea arises from the great library of ideas contained within the vast mind of the One Supreme Creator. The manifestation of that idea is imagined. One pointed concentration on the manifestation desired is required before it can be made manifest. The mind must be stilled of all extraneous interference.

For most of humanity this process seems difficult. There are many distractions, especially for those who choose to live in polluted environments.

Those people who exercise power and control over the mass of humanity choose to live in huge estates surrounded by nature.

The plant kingdom creates abundantly. The natural process of creation can be observed by those with eyes to see. A plant knows its purpose and follows the steps to achieve it.

Through human intervention of genetic manipulation, the natural process is experiencing decay. Unhealthy mutations are occurring and will continue to do so.

Purity of thought is needed for the creation of a pure vibration. By connecting with, and appreciating nature, that purity of thought develops within the human mind.

A polluted and unnatural environment adversely affects human thought.

*

On a subsequent visit, each student was given a seed and advised we were unveiling the mysteries of the stellar system and the parts of our bodies it connected to. As instructed previously, we were asked to uncover the knowledge from within.

The central Sun, Sirius, can be identified as a living, pulsating heart, filled with Cosmic Light and Prime Creator's love. There are nine spokes of radiating streams of light and love energy emanating from this central heart that enter the nine other heart systems, or sub-stations. Each of these systems connects to a particular organ in the human body. The organs are those of perception. The intelligence of the heart rules these organs through the senses.

My homework, I was told, was to connect these organs to the specific systems and then report back to the class.

As I write, I often need to pause and move away from it. The Arcturian energy embedded into the transmissions is powerful. It may be the same for you. Take a break if you need to, and return to reading when you feel you have integrated the previous content. The content is profound, and is coded to activate dormant consciousness particles that enable your remembrance of soul truth.

*

Love Infuses all it Touches

I travel to Arcturus and experience the following scene in front of my eyes: The pulsating heart of Sirius is bright and alive, emanating the power of love. The energy is palpable and my heart rejoices as I view and feel it. I understand that, as more and more people adapt the truth of this system into their lives, greater love will flow into their hearts.

It is the love flow that has been severely tampered with to make way for the supremacy of the intellect. This is the outcome of the cover-up.

Now there are at least twelve people who are aware of this, and five are willing to expose the cover up so the love flow can again take its natural place in the heart intelligence of humans. It is time for the collective serpents of wisdom to gather force, to shed their old skins and reveal the formerly hidden layers of truth stored in their subconscious. The truth shall set us free.

The pulsating heart contains all knowledge, and the wisdom of The One Source, which is Pure Love. It is the generating force, the motor that drives humanity and all kingdoms of nature. Love infuses all it touches. Love makes the world go around. It is from Love that we came, and it is to Love we return.

As I view the downloading of love from the Sirius pulsating heart I see the other involved star systems fill with the love light. They pulse and radiate love within their field of influence. I feel the process in my heart.

In esoteric astrology, the seven colour rays permeate our zodiac through three great constellations, those of the Great Bear, Sirius and the Pleiades. It is Sirius that holds the greatest download collected and received by an even greater star. This star is Arcturus, now the foci for beaming light to others.

The Sirians know the great responsibility they have been given to shine their light brightly to all on earth. Sirius is the brightest star in our night sky. It seems amazing that humanity bought into the cover-up when the evidence of truth is in front of their eyes. Venus, the planet of love, and Sirius are the two most prominent cosmic bodies in the night sky. Wake up people.

*

Love Heals All

I was asked to record and share the following transmissions while still immersed in the Arcturian frequency bubble. This bubble feels like bliss. My mind is alert and still, my senses fully active, love fills my heart and I feel like a small child perceiving reality differently.

The transmission adds further understanding to the story.

Love heals all. Love is the connecting filament of light that permeates all creation. Creation occurred through an act of love and it is to this love all will return. Love is the highest vibration and changes the structure of matter. Matter is a coagulation of dense energy vibrating at a slow speed. The energy of love vibrates at a high speed.

A high frequency stimulates and energises coagulated matter, piercing it as if a ripple on a pond. Waves form on the pond that change its structure. When the love penetration is great, the ripples in the pond are great and create even greater disturbance to the structure.

Electricity is generated from electrons. Photon energy is faster, and is pure. It comes from fine particles of photonic substance that are carriers of high frequency light. A photon band encircles your solar system emanating from a more refined area of space. These photons act as nerves that contain consciousness.

Consciousness is the fabric of the universe and is based on love. These photon waves are increasing in intensity and will make larger and larger waves upon the pond of the collective consciousness.

For those stuck in the material world of dense matter, a great awakening is imminent. Quantum physics has identified photons and the waves of photonic light. There is still the connection to consciousness to be made, but this will occur soon.

The great wake-up call to love is resounding throughout the cosmos. Humans have experienced the pain of separation from love, and this pain is the cause of all wars and disease upon this planet.

Love is the miracle of life. Love can be generated consciously through choice, intention, will and attention.

*

The Mind and Sacred Geometry

During meditation, without thought or expectation I wait at my 'Way Station'. Telepathically, I'm asked to open my wings and to fly. It took many years of practice to develop my telepathic abilities and etheric wings. My first journeys into space are written about in *I Am an Experiment, an Extraordinary Spiritual Adventure.*

As I take off heading into unknown space I experience a comfortable black void of nothingness. It does not feel dense or heavy. Far ahead, I see a faint light, and fly to it. I hear, and feel sound as a low humming vibration.

Caught into the waves of sound, I become engulfed in the associated Cosmic Light. The consistency of the dark and the light feel the same. The sound vibration, and the light illumination make the difference. It is a pleasant experience. I bathe in the light. Quickly I am guided to the Arcturus Board room table, and take my seat at the round table. No one else is present.

I look above to see beautiful light illuminating the table and then look down to see the white sheet covering up the true zodiac on the table. I hear my name, and am told I am correct in my intuitive understanding that Sirius rules the higher mental functions. I thought maybe it could be two parts of the brain, one being the arachnoid membrane and the other the hippocampus. I was told the others would be given to me. Hmm! I was asked to meditate each day to receive the information.

During meditation the following day I was advised the mind was created through sacred geometry and through the science of triangles. I could sense and feel this to be true. However, I asked myself, what is mind?

Is it the total all enveloping body's intelligence that comprises the etheric field around the entire body? Is it the brain? Is it both? Or, is it consciousness? I sense it is all. Whatever is above is below, in every form. The triangles are above in the cosmos, so they must also be in the body's etheric structure, and in every cell.

In the beginning of time before the spark of mind was implanted, humans created according to their natural instincts. There was no thought attached to their creations. Manifestation was feeling based. Feelings aligned humans to their natural environment. Humans were an extension of nature. They felt each other.

This evolutionary process existed for astrological ages of time. Natural harmony was experienced. At a pre-determined time, according to the Prime Creator's plan to know itself, the spark of mind was implanted into the human species. This gave the species the opportunity to objectify and individualise. A switch, once dormant, was activated. This switch resides in the frontal lobe of the brain and is almost mid-centre. This switch can be turned on or off consciously.

Following this transmission, I asked the question, "What star system activates the switch?" The answer was *None – it is a galactic synchronisation button, activated when consciousness is ready to utilise it.* I experienced this button to be a magic formula, or elixir, that immediately changes a state of being.

*

Connecting Images

Overwhelmed, I wanted to make sense of all I had been taught, but didn't have the consciousness to do so. During the years that followed four repeated images were given to me, related to the Arcturus Astrology system. I guess they were given to help me on the way. The first one had different parts to it, was repeated on five occasions, and mentioned in detail in *The Magdalen Codes* book.

1) Three Suns in alignment as:
   a. Our Solar System Sun
   b. The central Sun of the Pleiadian system around which our Solar System orbits; and
   c. The Sirius Sun, the great central Sun of our new cosmic playground.

2) A circle image of ten living, pulsating hearts. The larger single Sirius heart is situated in the centre. The other nine hearts were connected as one through nine spokes to the central pulsating heart.

3) The three Suns and the ten hearts were being influenced from above by a huge column of soft, feminine 'Goddess' light. This is from Arcturus.

4) A series of red triangles.

The second symbol above is depicted on the cover of *The Magdalen Codes* book, symbolised by a ten-pointed star.

*

Spiritual Ascension

Let us begin.

Planet Earth was created as a home, a haven for humans. The plant kingdom was created for the benefit of humanity, to assist their growth and healing. The plant kingdom was created to serve humankind. Many plants have healing qualities. Some heal the physical body and some heal the soul. However, not many people choose to consider this, and very few stop their busy daily routines to smell the roses.

The earth absorbs human density. Whenever the density becomes too great a burden, the earth expels it in the form of volcanic eruptions or earthquakes. Density is the energy of negative thoughts and emotions. Not only is this energy absorbed by the earth and felt by the animals, but it also is carried into the atmosphere, where it gathers mass. It becomes as if an invisible cloud. Where there is an accumulation of human density there is greater atmospheric pollution.

Thoughts and emotions are energy. Energy cannot be destroyed but it can be transmuted. Higher vibratory thoughts and feelings transmute density. Love and pure thoughts of good create higher vibratory patterns. This can be successfully proven. Therefore, if people choose to consciously create thoughts and feelings of love and goodness, this energy creates a power force that pierces and transmutes dense energy patterns.

The feelings of love can be consciously generated. It is easier to generate these feelings when surrounded by the beauty of nature, however it can be done under any circumstance. All it takes is concentration, focus, inner stillness and the willingness to do so.

When large numbers of people gather together to create peace, and they consciously generate peace in their hearts, their collective energy goes forth to create peace. It is when human thoughts and feelings are aligned in common purpose that positive results are obtained.

*

In ancient Egypt, initiates were trained in the pyramids to experience the ascension process. The levels of training were many and varied. Many tools were used to aid the process. One of these tools was the science of energy.

This was the astrological system that had Sirius as the central star. Life was lived according to the inundation of the river Nile which was determined by the apparent movement of Sirius. It was understood that Sirius was the great central Sun beyond your Sun and, as the brightest star in the night sky, containing and emanating the greatest light. Light is consciousness. This system of astrology was based upon the Sun of Suns. (Son of Suns?)

Sirius was known as the source of power for the greater universe and it was on Sirius that many of the great spiritual teachers were birthed. The initiates in the pyramids would be trained on Sirius during their meditations, and it was as a result of their direct experience that knowledge was gained.

The ancient Egyptian teachings are as relevant today as they were in those times.

*

The Divine Feminine and Masculine

It was when Yeshua and Mary Magdalen began their ministry that the knowledge of the ancient and remembered astrology system, the mother of all systems, was re-awakened. There are many references to astrological alignments in the old texts and secret writings, however the system we are discussing was not interpreted because it was not known by translators.

It was the system studied and revered by the ancient Egyptians, as they were taught it by the Sirians. Mary was a priestess of Isis and Yeshua ben Joseph trained in the esoteric mysteries of Egypt. Both were highly trained initiates of royal bloodlines, and were keepers of the sacred knowledge.

Not only was the Sirian system covered up but the knowledge taught by Yeshua and Mary Magdalen was also covered up by those who ruled.

It is time for the truth to be known. The truth is not discovered in the halls of academia, nor is it discovered in texts written second or third hand. It is found in the cellular memory banks of those who know. To access this memory, one needs to develop sensory perception, and allow it to guide your life.

Under the influence of the photon band, the huge band of photonic light that permeates the human brain during the Age of Aquarius, this memory is being awakened. Many will wake up and recall the truth. Not all will write it. This book will trigger many to remember.

*

PART THREE

The Eagle Flies High

To many indigenous people, the Eagle, and its South American counterpart, the Condor, symbolise the expansiveness of Great Spirit. Like these two great birds, we too can connect to our inner spirit of expansiveness, flying high in our consciousness to experience the world of Divine Light and Love while living on planet Earth. We can become super-conscious human beings.

Reflection and De-coding

The passages in italics that follow are relative excerpts from the Arcturians and The Tibetan, Ascended Master Djwhal Kuhl (D.K.), spokesperson for The Record Keepers. Many more of their transmission can be viewed on my website. www.ashtara.com

Following is a message from The Record Keepers.

As in times past, and in times to come, you will realise the value of our teaching. It has not changed. A great Plan for human evolution is in process. Having experienced existence in the pure Light and Love of Prime Creator your adventure into selfhood began. Descending into other dimensions, each denser than the other, eventually becoming enveloped by the third dimensional frequencies has been your chosen experience. Much has been learned and accordingly recorded. For many humans, the process of ascending has begun.

As each purifying step is taken to release the hold of dense psychology, an acceleration into finer frequencies of cosmic Light takes place. When the mind is trained to release the thought patterns that create emotional distress, greater light enters the space where psychological density formerly existed. The desire to purify one's inner space is needed to enable the light producing process.

When human minds are focused on external events and other people's life movies, where is the space for self-reflection, meditation, and prayer?

You choose your personal pathway of return to full enlightenment, at one with Prime Creator. You can allow material world distractions to hold you captive in third dimensional density, or choose to discipline yourself to daily self-reflection, meditation, and thanksgiving prayer. One is a path of shadows, the other the path of light, love and joy.

Your soul is connected to the path of light, love and joy. By listening to, and following its heart-felt guidance, your return journey to union with the One Source can be accomplished with ease and grace.

We leave you now to ponder upon our words.

*

When about ten years old, I looked into the bathroom mirror. Looking back at me with kind and compassionate eyes was a familiar face, completely different from my own. A strange face, with large almond shaped eyes, clear skin and pointy ears.

I was shocked. I didn't tell anyone because of my fear of being ridiculed. No longer did I listen to my inner voice. I stopped noticing the higher frequency awareness, the natural guidance every child has imbedded within their heart, soul, and mind.

About thirty years later I saw the face again, in the movie Star Wars. Again, I kept silent.

The ability to access higher frequency awareness and thought can be reactivated. Many people are now doing so. Opening to the living information of the life force rising within is the only guidance needed. We are not separate from the Source of life, nor are we disconnected from Supreme Intelligence who, designing all form and structure, breathes into it the breath of life.

We are the prototype of a new, universal species, a new race of humans who will span the gulf between the visible and the invisible. Through us, a new and unprecedented cycle of creation will occur. As a reincarnating soul, experiencing a gestation in the invisible realms before entering the stage of the visible third-dimensional world of form, so too does light, sound, galaxies and all they contain experience a gestation period. A continually replicating creation system unfolds according to immutable Universal Laws.

Into the Supreme Mind of the One Source emerged an idea, to create aspects of Self so it could know Self. In the beginning was the word, a sound carrying a frequency. Some say this sound was AOM. Emanating from the idea and sound of Source is a simple vibrational energy system of cosmic management, replicated and true in all cases.

Based on Divine Creator's love, this system of sound, light and colour permeates all of creation. Forming patterns, the always in motion sound travels on frequency wave bands creating vibrations according to whatever it encounters. Vertical light beams of colour Rays permeate the horizontal sound, creating geometric forms and shapes.

The Divine idea travelled, manifesting accordingly. Prime Creator thought of playmates, and the Elohim creator gods emerged to serve Prime Creator's idea and focused thought. Energy of love without form, they are filled with the purity, radiance and love of Infinite Supreme Intelligence. Glorious is the magnificence of creation.

Enjoying the light, stars and vastness of space itself, the wonderment of creation worlds continues forming as the Supreme idea permeates all it encounters. Archangels and angels emerged, filled with Prime Creator's essence of love, creative intelligence and consciousness.

Beyond all time and space, the pure energy of immense love continued its exploration.

Through the Universal Law of Attraction, magnetic particles formed as self-replicating atoms of differing sizes and shapes. Radiant stars attracted those of like energy and star systems formed, according to vibratory levels. The frequencies omitted created the scaffolding for orbits around the brightest star. Universal Intelligence orchestrates a huge galactic network.

Planets formed, each with their own individual packages of energy and sparks of Divine Light. Creation continued as Prime Creator's original idea travelled further. Unique and amazing species formed, each containing the love and consciousness of the original idea. All aspects of creation contain the original idea, to create aspects of Self in order to know Self.

Pro-creation developed. Human and animal life emerged on planet Earth.

When does creation end? Does it ever end? I think not.

*

A clarion call is resounding throughout the galaxy. Humans are at it again! Will Earth have to suffer again? This call began when a human first walked upon a different cosmic body from that of Earth. The Moon. And now, we want to mine the Moon, Mars and whatever else we can.

We still cannot exist as a species in a state of cooperation, consideration, love and grace. We still cannot see our neighbour as an out-picturing aspect of self. We still haven't become aware of, and purified, the psychological cause of human problems, the contents of our subconscious minds. We still don't know ourselves as body, mind and spirit. We still haven't realised we contain within our bodies and minds the pure essence of Prime Creator's love.

Humans have existed for millions of years. According to esoteric science, we have destroyed four races and are in the process of destroying our current one, the Aryan race. Have we learned to know ourselves in order to know God? Have we learned we are an organ of consciousness within a single infinite Being, as is our precious planet? As is all of creation.

Arcturus Astrology will accelerate your consciousness. Many changes to your perceptions and perspectives will take place. My focus is on guiding you into higher or Christ consciousness, so you can re-align with your soul's original intention.

As I continue my story, you will have opportunities to remember your soul's empowered design, and to receive internal guidance. This new cosmology of consciousness prepares you for union between spiritual and scientific understanding that must converge in order to make a quantum leap forward into the New Age of Aquarius, the Age of Light.

Nestle in while we decode together the unveiling of broader fields of awareness. Invite your analytical mind to rest as you encourage your intuitive mind to adventure into limitless fields of Light. Allow your heart and feelings guide you to truth.

As the Earth moves deeper into a different area of space, we humans upon her have the opportunity to move into a different consciousness system. The Arcturus Astrology System applies to the collective advancement of the human race, rather than to the individual. It's a system of consciousness by which the intellect must bow to the spiritualised intuitive higher mind.

Divine Love is the guiding force. To know oneself as Divine Love is the ultimate purpose of life.

The Record Keepers have the following to say:

We will not leave you now we have a scribe who is willing and able to hear and record our messages accurately. Much is still to be revealed. Little by little, as we see you are able to comprehend, we will add more.

You, as a race of humans have existed in slightly different forms for millions of years. On a few occasions in the past, your collective mind destroyed your civilisation. Some survived. Your human race continues. You are on the brink of doing so again.

Our scribe has memories of observing a previous civilisation, Atlantis, destroying itself. At that time, she made a vow to do whatever she could to prevent the same happening again. We are here to assist her. She observes a repetition of the same behaviour and collective mind set.

If you could only open your eyes to see from a higher perspective than your current purely materialistic one, you would realise with wonder the magnificence of your creation and of how you create your reality by your thoughts and subsequent actions. There is no need to create drama or psychological pollution.

Yes, you have karma to work through. The Universal Law of Cause and Effect is always in motion.

You can create loving karma through sincere good will and loving kindness. Dramas will then cease because you will perceive life differently, from a higher perspective. We suggest you learn to view yourself and your thoughts and feelings from a watcher in a tower perspective. All it takes to do so is intent and practice. You will then realise how you create accordingly.

You are magnificent creators, but mostly unconscious ones. Sleeping under a heavy blanket of mental and spiritual amnesia, we seek to awaken you. Each awakened one contributes to the collective mind, adding a spark of light. The light glows within the darkness and joins with other sparks to create even greater light beams. Like torch light shining into the night sky your light beam brightens assisting others to remove their amnesia blankets.

You can choose to be a bringer of light.

Many pockets of light are seen by those with eyes to do so. We rejoice.

To stabilise your planet, as it hurtles through unknown areas of space, requires many more sleeping ones to awaken to self-awareness. Know Thyself so you can know God. The essence of God, in the form of higher consciousness - which is Light based on love, is within each of you.

Light and love are your essence. You cannot be separated from your Creator. You can think you are.

*

The time is fast approaching when the prophesised changes are to occur. All it takes to make the internal change is to realise you are a part of God, of Prime Creator, and cannot be separated.

Give yourselves quiet time and space in your daily lives to allow this truth to anchor in your hearts and minds. Bask in the warm feelings that will follow from the acceptance of it. When anchored and you open to the warm feelings of love, your lives will change for the better.

This simple act will help in stabilising planet Earth. She, like a mother, nurtures and provides for you. You can show your appreciation of her through the realisation you are both aspects of the One Prime Creator, existing in an infinite field of love.

We suggest you do this practice regularly, on a daily basis, until you feel the joy and unification in your heart and soul. You will then know you are loved beyond measure.

*

As it was two thousand years ago so it will be again. A Christ was born to demonstrate the ability of a human to purify psychology and to resurrect into a body of Light. Jesus said that whatever he did you can also do.

Every human soul is connected to the One Source. No matter how much psychological baggage each one has built up over many lifetimes, the purity of that connection is there. You can choose to consciously work through dense polluting baggage, allowing Soul's light to shine more brightly, or not.

For those who do, a new evolution awaits. You will leave behind all that prevents access to the internal Christ Light, and you will radiate good health, happiness and joy accordingly. This state of being love, and allowing Soul's Light to shine clearly, is referred to in the old sacred texts as the second coming of Christ.

It is prophesised, and is in the making.

We leave you now to ponder upon our words.

*

The Great Work

In order to make practical sense of the Arcturus Astrology system, the following is an overview of foundational information. I'll briefly itemise because the information is written in greater detail in my two previous books of this trilogy. But first, a reminder of my persistent enquiry.

What is the cosmic system that activates further human evolutionary spiritual growth when we have worked through the psychological issues depicted in our birth charts?

Arcturus Astrology is the answer.

There is a Great Plan for human evolution devised by a Supreme Intelligence (Mother/Father God) consisting of Love, Light, consciousness, mathematics, physics, sacred geometry, sound and colour. The use of spirals, triangles, pyramids, merkaba, and other geometric structures, are but a few involved in the Grand Design.

Frequencies of Light and sound create sacred geometry as the Tibetan monks frequently and easily demonstrate through the creation of their sand mandalas. As the monks unify through love, and chant the sacred sound of Om, the sound vibrations create different patterns and shapes within the sand mandala. The monks are able to read the meaning of the mandala and share it with those interested.

We humans are in the process of activating our personal spiritualised Light from within, as the Ascended Masters have done. When we have developed sufficient Light, an evolutionary process now being accelerated, we will vibrate accordingly.

We are assisted in this developmental process by the planets in our solar system. Solar Systems are stepped down into different frequency zones according to their levels of light, yet all are connected to the One Creative Source. The work of the planets in our Solar System is to reactivate into our awareness the human soul's original understanding, wisdom and knowledge.

Likewise, our souls descended from Source into differing frequency bands to experience life in the lower realms, gradually forgetting connection to Source. Ascension into the higher realms of Light frequencies is a journey of remembrance and the path all humans walk, no matter how many lifetimes it takes.

This spiritual evolutionary process is being accelerated during the Age of Aquarius. We will eventually, as a collective, arrive at the conclusion we are all One, living within Prime Creator's Source Field. "As above, so below. As within, so without".

The Record Keepers share more enlightening insights.

*

Time is an illusion, a mental thought-form purposefully projected into human consciousness as a means of control. Yes, as Earth travels around the Sun day and night are encountered. Day – or light- symbolises consciousness, night symbolises the unconscious.

To become super-conscious human beings requires many forays into the unconscious, or hidden realms of the human psyche, to reveal subconscious motivations. When revealed and understood, light exists where once there was darkness.

The process of Earth's journey around the Sun is cyclic, as it is with the other planets in their movement around the Sun. The further out from the Sun is the planet, the longer is its cycle. Each planet emits a certain vibration, a sound wave, the frequency of which is experienced in the human psyche.

Through the study of astrology, these effects can be made conscious.

Our scribe learned to feel these effects and connect them to psychological problems via the individual's birth chart, or horoscope. This esoteric science has been known and practiced for thousands of Earth years. Indigenous races have their own versions of this natural phenomena. Each culture has its own interpretation.

The human psyche experiences the energetic emanations of the cosmos. Humans can learn to feel these movements in the brain. This practice enables the conscious mind to awaken to the information contained within the subconscious realms. Illumination occurs. The "lights" come on in the individual's psyche. Self-realisations occur, enabling comprehension of psychological density. This process requires follow through action to enable the inner light to permanently fill the space where darkness formerly existed.

We offer this simple explanation to aid comprehension of our messages. We leave you to ponder upon our words.

*

In times to come, when humans are more awake and aware, greater peace will reign upon earth. You will realise it is created from within as a result of collective thought and behaviour. The process has begun and it will be like a tidal wave, clearing psychological debris within its path.

Be prepared for this purifying process to happen in your personal lives. The wave has begun and will not stop. Understanding that the end result will be beneficial will help you to ride the wave with confidence.

In astrological terms, transiting Pluto moving into the last few degrees of Capricorn, along with transiting Uranus in Taurus, (two earth signs) will initiate the transformational changes needed.

As mentioned previously, the Great Plan for human evolution into the light of higher consciousness is in acceleration mode. The choices you make individually, whether they are unconscious and re-active, or conscious and well considered, will constitute the results experienced.

When aware of the consequences of each choice made, knowing it will affect your life, you make them wisely.

We leave you now to ponder upon our words.

*

Energy Awareness

The Arcturians share their perspective:

Everyone has inner access to the information circulating in the unified field of awareness that precedes all individuality. We have access to Prime Creators Infinite mind. Love is the key. Love of the internal Divine Creator existing within every cell of our bodies.

The revealing of Arcturus Astrology is an attempt to awaken you not only to the predicament you have created through unawareness and a perceived separation from our connection to the Divine Source of love, but also to a cosmic system available to those willing to do the inner reconnective work. When it is fully realised we are all a spark of Divine Source, and have the ability to create life through love, we will avoid the human abuse and calamities planet Earth previously experienced in ages past.

A cosmic system is in place to activate and perpetuate Prime Creator's original idea. The system is based on Divine Love. Nine cosmic hearts pulsate frequencies according to the one central heart radiating the pure emanations of Prime Creator's Love. This energy is readily accessible to the millions of people who, with love in their hearts who have done, or are in the process of doing the inner work to cleanse and purify subconscious psychology.

The science of energy, of consciousness, is a spiritual science. Life force energy, chi or prana as it is known in the east, is all around us permeating all of matter, and has a wide variety of different vibrational qualities. We can learn to measure, identify and consciously utilise this energy for good, or evil.

Many of the children born from the year 2000 onwards have this energy awareness. Unfortunately, most parents are unaware of it. These children can teach us much about energy, and about communication with the invisible world. Wise and conscious parental guidance enables this process.

Energy awareness is the feeling, sensing and knowing of invisible energy. It comes from using our higher perceptive faculties, our extra sensory perception. It was this energy awareness that enabled me to receive the Arcturus Astrology teaching. The experiences and teachings I received on different dimensional star systems are symbolic metaphors, connected to the material I was taught.

The Record Keepers have the following to say:

As in times gone by, and in times to come, many more people will hear the call of Spirit. This inner call may be faint at first, felt as a longing, or a yearning, for a something. Even though lives will be lived fully, they feel empty. Something is missing.

How does one unite with Spirit?

The process differs with each human being.

Essentially, providing quiet time each day just for self is the first step. Letting go of thoughts during this quiet time is the second. Being in Nature provides solace for the soul and it is easier in Nature to calm the mind.

Committing to doing whatever is necessary to claim your daily quiet "being" time is necessary. This simple daily practice will fuel your soul and provide the quiet space needed to receive your Spirit's guidance. Daily prayers of heart-felt gratitude for all you have, and have learned about yourself, will contribute to the process, as will daily meditation.

An unfulfilling life creates illness. Only you can change your life to one of happiness and joy. These state of being are your birth right. It takes serious intent and daily practice to attain a state of grace.

We leave you now to ponder upon our words.

*

To some, our messages will have little meaning, falling on deaf ears. To others, they will touch their souls and truth will be felt. It is as It Is.

As the blanket of amnesia is thrown aside more people will awaken and feel the truth within. We persevere.

All humans are at different levels of consciousness. To find others at a level similar to your own can feel as if you are reuniting with a family member, no matter the age difference. You are attracted by each other's inner light.

There is so much love, joy and wonderment yet to be discovered by those pushing aside their amnesia blankets. When the heart is returned to its rightful place in union with Prime Creator, the world shines with radiance and beauty is experienced daily. Joy abounds.

This state of being love awaits all humans willing to move through and beyond dense psychology. This love is found within and is different to the love felt for another human. Divine love rises to the heart centre as the understanding arises from within of the Eternal Presence. This spiritual experience of ignition is felt as a wash of warmth throughout the body and in every fibre of being and can be seen through inner eyes as tiny sparks lighting up.

This is the next human evolutionary step. Those awakening early are the forerunners of a movement that can only accelerate. Part of the Divine Plan for human evolution, the mass awakening cannot be stopped. Many will try and do so. Where there is light there is also dark. The materialising influence can be seductive and addictive. Human minds can become so distracted by it stress and disharmony abounds.

When decisions are made to follow the path of love and joy, stress and disharmony vanish. It takes trust to follow this path. Trust comes through experience.

Love and joy will never let you down.

We leave you now, and will come again.

*

The Zodiac

The signs of the zodiac symbolise twelve avenues of human psychological expression, involving six polarities. These polarities, when imbalanced, contribute to the duality and density existing in our third dimensional (3D) world.

To experience internal balance, the six zodiacal polarities require understanding and integration of their inherent psychology, and the willingness to work towards their optimum expression. When harmonised through deep inner soul work, and many revealing self-realisations, the light of higher, or super consciousness emerges. The gift associated with facing and resolving our mental and emotional pain is internal peace and joy.

Volume One of my trilogy of astrology work books, *Your Recipe for Empowerment through Spiritual Astrology,* is an empowering tool to enable students to optimise the positive psychological expression of each of the signs of the zodiac.

The energy of each sign operates within a field of diversity, yet all are connected by the wheel of life. All life is interconnected. What damages one, damages all. Perception is always the method for converting pain to empowerment.

Energy is required for conscious evolution. Shock, confusion and self-doubt drains energy. When we learn to appreciate, value and pay attention to our internal zodiacal archetypes, we notice each of their specific functional and dysfunctional psychological patterns being played out by those around us. The next step we learn is to view all of life as a mirror, guiding each step we take.

Because the zodiacal signs connect us to the cosmos, they are continually receiving energetic downloads about the known and unknown. Each sign naturally communicates their received information and insights to the others. Ignorance of our internal archetypes causes disharmony that can lead to health problems and disease. Each sign of the zodiac governs a specific body part.

When we refuse to move forward in our life our hips and thighs become problematic, and we can have problems with our sciatic nerve. When we have relationship problems and feel fearful of speaking our truth, throat and thyroid problems arise.

When we live life in fear, unwilling to develop self-discipline and take personal responsibility for our life creations, actions and psychology, our bones weaken. When unwilling to address and heal separation from love issues, we can suffer heart attacks. When we lie to ourselves to void facing issues, we have teeth problems.

It is wise to learn about our inner zodiacal archetypes so we can create healthy minds and healthy bodies.

As previously mentioned, the four elements within the human body that are essential for all life also need to be consciously harmonised to enable positive psychological expression. The way to do this is also described in the above-mentioned trilogy.

Each sign of the zodiac connects to one of the four elements, the building blocks of nature. Aries, Leo and Sagittarius are fire signs governing our spiritual body. Taurus, Virgo and Capricorn are earth signs, governing our physical body and practical connection to Earth.

Gemini, Libra and Aquarius are air signs governing our mental body. Cancer, Scorpio and Pisces are water signs, governing our emotional body.

Astrology is a pathway to Christ/Buddha consciousness and can aid our soul's ascension from this dimensional earth plane. It was, and still is, practised by the keepers of ancient esoteric wisdom.

The Church of Rome dictated it be abolished because it provided the means by which individuals could gain personal sovereignty and mastery. The Church created a hierarchical religious system instead, under which the attainment of self-realisation, or God realisation, could only be experienced through a religious intermediary. It took nearly two thousand years before this masculine power and control abuse began to be relinquished.

Intellectual technology destroyed the civilisation of Atlantis thousands of years ago and humans lost their higher levels of consciousness as a result of the shock of this catastrophic event. My first book transmitted by the Arcturians *Gaia, Our Precious Planet* depicts this fall in graphic detail. Do we want to repeat this same scenario?

We will, unless we as a collective "wake-up" to the truth buried within our subconscious, our personal psychology.

Without consciously bringing Source/Spirit/Light into our daily life we will continue to recreate psychological density unconsciously.

According to *Arcturus Astrology*, the current design of the zodiacal wheel of life was created following the fall of Atlantis, a period of history when human consciousness devolved. This downward spiral of devolution into ignorance of Divine Light and Divine Love is still in effect as is evident in our external world. This world is but a reflection of the internal world of the masses.

The good news is that this collective state of being human is changing rapidly as more of us awaken to a different reality, a reality involving the Light of higher consciousness and the love of our soul's Divine parentage. As we do the deep and penetrating inner soul work, we activate a sequence of higher-level multi-dimensional DNA codes. These codes work together to initiate the spiritual evolutionary process.

The process involves working daily with Spirit to enable openness, self-realisation and awareness. This enables activation of our natural extra sensory perceptions. Spirit/Light harmonises and integrates the keys to our multi-dimensional DNA, that when decoded, activate our Christ, or higher consciousness. Multi-dimensional DNA reflects the qualities of the soul. Our physical DNA reflects our genetic heritage.

We humans are earning the right to reactivate our original DNA strands that were tampered with millennia ago. We have help from some of the extra-terrestrial beings who consciously did the tampering, mostly with good intentions. Some of them were the Pleiadians, Sirians, Orions and Arcturians. Sedna is another. These beings admit they have interference karma to repay and are doing what they can to do by aiding our consciousness raising journey.

Arcturus Astrology, a consciousness raising system for the Age of Aquarius, is a journey that can lead to our soul's ascension from continuous rounds of earthly incarnation. We are becoming galactic, or multi-dimensional human beings, and our Souls are returning to the One Source from which we came. In simple spiritual terms, Great Spirit/Source is calling Its chicks back home.

The Arcturians say:

The trinary central star Sirius (A, B and C) represents the God Presence, and the spark is the seed containing the knowledge of All That Is. When this seed is nurtured with love, respect and conscious awareness, it sprouts into a healthy plant. Knowledge, wisdom and strength is its branches. One knows, without knowing how one knows.

With further nurturing, care and consideration, the plant becomes stronger. Its growth radiates to different parts of the body, and fuels those parts with its essence. The greater the love and nurturing, the healthier are those parts. It has always been so.

In ancient times, people lived for hundreds of years. A gradual shortening of life took place in humanity's long history, in direct relationship to self-love, self-nurturing, and honouring of the internal Divine Presence. When humanity began to worship idols and things outside themselves, human life span shortened.

Now that humanity is again beginning to honour, respect, love and nurture their inner God Presence, their life spans will increase. They will gain more life-force energy and subsequently live longer lives. And, so it is, and always has been.

The Solar System Planets

The work of our Solar System planets is to energetically assist humans evolve into higher frequencies of light. We can do this consciously by following and working with planetary transits to our birth charts.

When we succeed in purifying our subconscious psychology and attain a higher level of consciousness our Solar System planets will have done their work. When this higher level of conscious is maintained as a constant the Arcturus Astrology system comes into effect. The following guidance provides examples.

Mercury: rules the signs of Gemini and Virgo.

In service to the gods, Mercury is their trusted messenger. He listens to their messages and delivers their content. We have the opportunity to listen carefully to our quiet internal messages from our Higher Spirit Self and Soul, or not.

We are also messengers, in service to the Divine Plan for human consciousness evolution. Our soul's messages are received via extra sensory perception, intuition and heart felt feelings. As we evolve, these messages can come via telepathy through direct clairaudience. To evolve and grow into higher consciousness, we need to listen to, honour, respect and act upon these messages and share them where appropriate.

When we have learned to honestly communicate through our hearts, without blaming or judging others or ourselves, and learned to carefully listen to, and act upon our inner guidance, Mercury will have done its planetary work.

Venus: rules the signs of Taurus and Libra.

Venus, the planet of love, beauty and feminine power, in her cosmic travels forms a –five-pointed star, a pentagram, over a synodic period of approximately eight years and five days. In ancient Egyptian times she was associated with the great goddess, represented in the night sky by Sirius, the so-called Dog Star situated in the constellation of Canis Major. Venus and her glyph play a major role in our emerging world of higher consciousness. Many souls spent incarnations upon Venus. I remember one.

In the bible, Yeshua said he came from the morning and evening star. Venus connects to our values. Yeshua demonstrated how much he valued women equally with men. The Church did not appreciate his view.

As we evolve into greater light, choosing to live according to our soul's heart-felt values and well-developed levels of self-worth and self-value, we contribute to collective peace and harmony. As we learn to speak our truth from our hearts, with love, Venus will have done her planetary work.

Mars: Mars, the planet of war rules the sign of Aries.

Mars symbolises our 'warring' nature, and what we will fight for. The area of life in which it is placed in your birth chart is where you will need to confront your 'enemies' (your psychological demons). Also, the placement shows to where and to what you are most likely to direct self-assertive energy.

Those who have done the deep inner psychological soul healing work are referred to by the Arcturians as the Rainbow Warriors. They have fought their way to the Light by being unafraid to confront whatever psychological darkness their souls have accumulated over many lifetimes. They will have faced and embraced the feeling truth revealed, no matter how devasting or dark it appears to be. They will, with full intent and bravery, alchemise the relevant psychological themes and issues still influencing their current lives.

When you have learned to become a spiritual warrior, standing up tall for what you feel and know to be true, and willingly live it, Mars will have done his work.

Jupiter, the expansion influence, rules Sagittarius.

You will have encouraged Jupiter to open your higher mind to understand broader, wider and higher spiritual concepts and perspectives. You will have learned to realise, speak and live your truth, and have willingly discarded controlling religious doctrine.

You now allow your intuition (inner tuition) to guide you through life, and are open to, and have faith in, spiritual guidance from higher realms. You are willing to continually grow and expand your consciousness.

Realising you have been accustomed to looking outside yourself for guidance and reassurance from an external source, you now realise all truth is within, and is relative to the beholder. You trust your inner knowing and act upon your quiet voice of intuition. You reach out into new perspectives of truth, and, when you feel a warm and loving vibrational resonance within, you trust and act upon it. Jupiter will have done his work.

Saturn: Saturn rules Capricorn.

Saturn, the materialising influence, is the principle of contraction. Until we work through our historical karma we can't move on to experience higher octaves of life experience. Fear of the consequences of moving forward hide in our subconscious.

Contraction and currents of fear can be experienced as hell, shutting off eternal awareness and the energy currents that bring love and beauty into the world. When you commit to working through your fear no matter what, or how long it takes, life will transform.

Those desirous of entering the new Light evolution understand that Saturn is the structure of consciousness. You will have learned to break through your psychological brick walls to realise that only you create your reality by your thoughts.

You do not blame, criticise or judge another because you know you are responsible for all you create. You have learned self-discipline and patience. Manifestation of ideas can happen almost immediately according to your ability to detach from the outcomes you create in your mind.

You will have chosen to break through your psychological defence barriers through self-love, self-value, self-reflection and self-realisation, and now take full responsibility for your life creations. You realise the consequences of every action and thought you make.

You are aware of, and know without a shadow of doubt, that the Law of Karma, of Cause and Effect, is ever operative. Whatever you cause by your thoughts, emotions and deeds has an effect, a consequence, no matter the life time of the cause. You have learned to become self-determined, self-reliant, and sovereign unto yourself. You are your own authority. Saturn has done his work.

Chiron: Chiron is said to co-rule Virgo.

You have opened, realised, accepted, and loved your psychological wounds to death, have been compassionate towards them and have consciously worked on alchemising them. You know they are healed when you no longer react to emotionally charged situations. No longer do you have any emotional charge to the person, or persons, who proved to be your best and most challenging teachers.

You have allowed Chiron to open the closed doors of your psyche to enable the downloading of spiritualised Light, the Light of higher consciousness. Through bridging the inner planets and working through the energies of the outer planets, you assist others to do the same. You have become your own healer, are wise to the ways of the world, and offer keys of healing wisdom to those who come to you. The work of Chiron is done.

Uranus: Uranus rules Aquarius.

You have encouraged, and asked for Uranus to bring flashes of self-realisation, those electrical light impulses that illuminate your thoughts. No longer do you wish to experience chaos caused by separation from Divine Love. You have moved forward into the greater Light of self-awareness and now have a completely different perspective of yourself and your evolutionary spiritual path.

You now view yourself and your life from a much higher perspective, understanding there are many ways of perceiving reality. You choose to walk the higher road towards your soul's ascension from this earth plane. Realising you are being constantly challenged to expand your awareness and reach out into new finer frequencies you willingly accept the direction of your inner Light without being concerned about where it will take you. You become the quiet observer of your inner self viewing life from a higher perspective. Uranus' work is done.

Neptune: Neptune rules Pisces.

Through feeling and intuition, you realise the true from the false, and are able to determine how the marketing manipulators use certain words and phrases to sell/feed you with what they've been told and paid to do. Through feeling, you will have realised, and accepted as true the ever-present availability of Divinity, and are open to receiving and sharing Divine love.

You have developed faith in self and your unique spiritual journey. Realising that Divine Spirit works from within guiding your way, you invoke as much Spiritual Light into your inner being as you can handle. You've learned to sacrifice your heaviness and density, fears, doubts, indecision and limitations, and have surrendered your personality angst to Divine Spirit.

Having developed compassion for yourself and all beings, you now know and accept you cannot rescue another human from themselves, their life decisions and creations. Nor are you willing to ever play the role of victim again.

You have learned to forgive yourself and others for the perceived harm experienced, realising you attracted it to further your self-awareness growth. You hear the call of Spirit, and act upon it. You have developed your extra-sensory perception and allow it to guide your journey to the Light. Neptune has done his work well.

Pluto: Pluto rules Scorpio.

You have made friends with, and alchemised, your dark psychology. By so doing, you are aware of old cultural, societal and parental conditioning that no longer serves you, and refuse to allow it to take you down into an abyss of negativity. You know that, unless you release old ways of perceiving yourself and life, you will continue to replay them experiencing the same results. You have become the observer of your thoughts and feelings and can now trace disempowering thoughts back to their source, and alchemise them through loving acceptance.

Realising that Cosmic Light provides the fire of transformation, and that the flames in which all that in not Light is consumed, you welcome and work with it via Pluto's transits to a birth chart placement. You choose to follow the Light of higher consciousness.

No longer afraid of the dark, the unknown, or death you know that with every positive spiritual and psychological step forward you receive the support and encouragement of your spiritual guides. They are always with you, encouraging your re-birth into greater love and Light. You know the Universal Law of Death and Re-birth and are unafraid.

You choose never again to descend into the darkness of psychological ignorance because maintaining your level of light vibration is far too important for the state of your mental and physical health, and the health of those around you. You no longer give your power away to another, or play the role of manipulative control and power over another knowing it could take many future life-times to heal. Pluto will have done his work.

Sedna: (Neptune's feminine counter-part) Sedna co-rules Pisces and Libra, with a dose of Scorpio.

Sedna was discovered in our Solar System in 2003. She is the bridge between 3D consciousness and that of 5D and beyond. Multi-dimensional consciousness is familiar. It was your state of being prior to your descent into the 3D world of matter. You can commit to reclaiming it. You may not have much time to do so. Throughout 2021 I received the same message: "The doors are closing". To wake up or wither seems to be the clarion call.

I'm reminded of my granddaughter's enlightened painting created when she was four and a half. The images painted were of our Sun, Earth, trees and many hugs and kisses. Explaining, she said to me she'd painted the hugs and kisses because she loves the Earth so much. She also explained another image, that of a second Earth. She is now an adult. As a small child she knew of this second Earth.

Should we not wake up to all that blocks the daily experience of Divine Love and Grace we will repeat endless rounds of 3D incarnations on a second Earth. Gaia will have ascended into a higher frequency band, and humans living life in the same frequency level will ascend with her.

There are many chapters about Sedna written in my *Magdalen Codes* book, given to me from first-hand experience. One of the stories relative to this book follows.

During a meditation when at my 8th chakra platform, I invoked Sedna and she asked me to go to where she was sitting on a large white quartz crystal, her body slightly blue, her mermaid tail slinky and smooth. I sat upon the large crystal beside her and noticed I too had the same tail.

She toned a high pitched "EEE" after using a tuning fork to tap on to the side of the crystal rock. She asked that I also tone the same sound. As I did, I observed the entire inside of my head feeling as if it were expanding, creating more space. My head felt like a sound chamber with the sound pushing against the walls to make room for greater inner space.

At one point I could feel my solar plexus chakra becoming energised. Then the energy moved to my heart. My stomach churned further as ancient memories stirred. I intend to uncover the information. Sedna suggested I open my Akashic Records book. I did, page 176.

And it came to pass that the joining together of two specific strands of genetic material resulted in a creature who could live under the sea and walk on the land. This experiment successfully led to another, slightly more complicated. A third piece of genetic material from a different species was needed and this species walked upon the land. The combination of the three strands of genetic material formed the base material structure of the current human race.

Inherently, within our geometric structure code is a triangle. The triangle is the basic geometric form of all manifestation, whether a solar system, zodiac, cosmic triplicities or human. We humans were created by three separate strands of genetic material originating from the stars. A combination of Pleiadian and Sirian material, and one from an existing Earth species.

The memory of this genetic splicing is stored in a specific gland in the human brain that connects to the stomach. The back of the head energises when the memory cell activates. Ancient Tahitian myths contain a small vestige of truth. The practice of phrenology also contains a connection. Research well.

As I read, I become aware of energy stirring in my stomach, (solar plexus chakra). Moving to my throat the energy activates further. My solar plexus, heart and throat chakras feel as if they're pounding. The vibrations are strong. The sound and the reading activated an ancient memory. I need to breathe it out through my mouth, as I had done many years ago when accessing ancient rage.

At that time the taste of the raw rage energy was dank, dark and foul. I saw, smelt, and sensed it to be black polluted smoke. Whatever the content of the memory I was now experiencing was a vast improvement. My earlier memory was based on rage against the collective masculine for sex and power games perpetuated against females over aeons of time.

"What is the basis of this memory?" I asked myself. "Genetic splicing? Interference by others to accelerate the human experiment?" I opened my mouth to scream, silently encouraging release of the ancient memories. My body responded. Grace filled the empty spaces.

The EEE sound vibrating in my head activated ancient memories enabling their release from captivity. The AHH outbreath was the release mechanism.

Sedna has come into our consciousness to rectify her karma associated with the Nephilim, the creator gods mentioned in the bible. She has come to help balance the scales between masculine Mars and feminine Venus by activating memories of our human origins.

From that diversion I'll continue.

You will have allowed Sedna to open you to divine love and to access multi-dimensional consciousness. Having developed mastery over the ancient soul memories that created oceans of emotion, you are now in charge of the 'creatures of the deep' who exist in your subconscious. Everything around is changing and evolving, because you are changing and evolving.

You are moving into a higher dimensional New World based on co-operation rather than competition. Oneness rather than separation. You naturally move towards all that is Light, light friends, light foods, light ideas and lighter ways of living. Sedna has done her work.

The work of the Solar System planets is done. You have practised deep inner work to raise your consciousness and are ready to move into, and exist within a higher frequency of Cosmic Light, a higher dimensional vibratory tone, while still living and loving on planet earth.

Together, as a unique aspect of a collective Light being, we have moved from the child stage of dependency on external authorities to the stage of spiritual empowerment. We are ready to experience and work with the group consciousness of Arcturus Astrology.

*

Sirius and the Council of Nine

Sirius operates at a 6D dimensional level forming a cosmic triangle to the Pleiades and Great Bear. Sirius is the great central Sun of the new astrology system that aids humans to develop galactic or Christ consciousness.

Over a period of many years the Sirian Council of Nine trained me to identify and to merge with each of the members, emphasising regularly that I am their only earthly representative. This information caused me to doubt my sanity. However, as their regular training proceeded, written about in my previous two books of this trilogy, I learned to accept their comments as a work in progress.

During one of the harmonising experiences with them, I became aware of how each individual member demonstrated the qualities we need in order to experience totality of beingness. When we embody these qualities, a merging with the unified field, often referred to as the Goddess, takes place. At one of my reunions the Sirians told me: *We are of the Goddess, and we are One*. Isis is the great goddess of the ancient Egyptian mystery schools and was sourced from Sirius, as was Ascended Master Jesus.

Omega, spokesperson for the Sirius Council of Nine, told me that each individual Council member guides, over-lights and oversees, through love and compassion, one of the nine satellites emanating from the central heart of Sirius. The unified goal of all members is to awaken and expand human consciousness evolution into greater love, and the light of multi-dimensional consciousness. Some refer to this attainment as God or Christ consciousness.

Each of the starry satellites in the system connect to a specific quality emanated by a Council member, and also to a particular star visited during my galactic journeys, written about in my two previous books. All the Council members I experienced emanated intelligence, understanding, discernment, love, wisdom, compassion, humility, trust and integrity. These were qualities that a group of Light Beings, The Tall Ones, told me many years ago they were assisting me to embody.

The Arcturus Astrology system, with its alive, pulsating central Sirian heart, is playing its part in Prime Creator's Great Plan for human evolution of consciousness. It is the medium through which greater love and light is being downloaded to all human life on Earth.

Each of the nine satellites in the Arcturus system receive the energetic emanations of specific qualities emanating from the Council of Nine from Sirius, the central radiating star. Omega appeared to me as a soft silvery colour, whereas RA emanated brilliant golden radiant light, like a bright summer Sun at noon.

The Council of Nine members, as they revealed themselves to me are:

Sirian Star Master: Omega
Feminine Buddha: Tara
Ascended Master: Sathya Sai Baba

Ascended Master: Yeshua ben Joseph
Ascended Master: Beinso Duenov
Sirian Star Being: Athenia
Human: Ashtara
Ascended Master: The Tibetan, Djwhal Khul (D.K.)
Sirian Star Being: RA

None of the Council Members demonstrated force or control.

The Sirius Council of Nine guide the nine starry satellites from the central heart of Sirius. Each council member over-lights and oversees one specific starry satellite. The nine starry satellites of the Sirius cosmic system I experienced and was asked to record, and the Council of Nine's qualities over-lighting each star follow:

*

The Nine Satellites, the Council Members and their Qualities

The Arcturians say:

The nine spokes of the wheel and the hearts they emanate from symbolise each member of the Council of Nine from Sirius. The characteristics and the qualities of each member connect to one of the stars you visited. You, Ashtara, chose to walk-in to an existing human female to bring this system into human awareness.

Connecting each player to a star will be easy. Simplicity, intuition and feeling are needed to crack the codes. You have access to the Earth's records. As one of the first wave of pioneers to visit Earth, your journey has been long, and often arduous. Yet you persisted. Your story-telling career is about to begin. We are looking forward to it. The Arcturus model is your tool.

Sedna is indeed important to your story. She is the connecting link to the galaxies beyond your Milky Way.

*

Antares

The Antarians trained me similarly to my final Arcturus experience when I was the only returning student. In both experiences, I was trained to unify all aspects of myself. The Antarians trained me to feel disturbances in my energy field; from all around me, front, back and sides and to feel their collective emotions such as fear, anger and love.

In other words, I learnt to read through feeling and sensing, their psychological energetic emanations. It was an exceedingly valuable training. Each one of them in the group demonstrated unity, group consciousness, joy, love, calm and compassion.

I allocated Sai Baba to Antares.
He was one of my early guides.
He laughed a lot and taught me to appreciate the great cosmic joke of life.
The main quality he taught me was Trust.

The Pleiades

The Pleiades are known as the seven sisters. They were early players in my celestial journeys. Often, with their advanced technology, they monitored my levels of consciousness and recorded them. I felt their love, respect, gentleness and gratitude. I travelled to the Pleiades and wrote about these journeys in my *Experiment* book. I viewed the Pleiadians as a happy group of little, faintly green, humpty-dumpty entities. Filled with the Divine Creator's love, joy and play, they demonstrated how their advanced technology acts as a consciousness frequency modulator. Alcyone, the main star of the system, is the second Sun star in the Arcturus Astrology system. This Sun beams radiance and love to our Solar System Sun. Our Solar System orbits around the Pleiadian System.

RA - Sun god

At all times I experienced him radiating so much golden light it almost obscured his slightly visible form. Focused on his task of self-generating and sharing Prime Creators' immense Light and Love the main qualities he demonstrated were unending, focused Radiant Love and Light.

Draco, the Celestial Dragon

A mythical creature appearing in ancient cultures. A former guardian of the feminine serpent, he kept the serpent goddess hostage, fell in love with her, and willingly relinquished the hold he'd had on her since enticing her from the Garden of Eden. His cosmic task had been to hold her as a sacred treasure in safe-keeping until the change of the astrological Ages. He did his job well. The serpent goddess is again capable of being her own power source.

Part of the Dragon's work was done. He ascended through an explosion of divine light where he can now be in service to the Great Plan in the new evolution. The ancient dragon/serpent feminine wisdom, embodied by Mary Magdalen, Isis, Quan Yin, Lady Meru and other feminine Spiritual Masters, can once again come into the forefront of human experience.

When I working with Saturn a few years ago, he morphed into a dragon, saying he was there to serve me. He said I had once rescued his feminine and had taken her into myself for safekeeping, and then directed him towards the Light. He told me he was repaying the favour as a cosmic dragon protector.

Tara, the feminine Buddha

Exuded grace, love, serenity and beauty. True feminine power. The main quality I perceived in her was serenity. She demonstrated natural Beauty.

The Great Bear

Awakens from a deep sleep into the role of cosmic herald, announcing the timing of the higher evolution of human consciousness. Glastonbury in the United Kingdom reflects the Great Bear constellation. It was, and probably still is, a spiritual initiation place for the Druids.

There is an ancient western style stone zodiac circling around Glastonbury that can be clearly seen from the air. It places Glastonbury in the area of Aquarius, represented by a phoenix, the mythical bird that rises from the ashes of its former self.

The story of King Arthur and his twelve knights of the round table connect to Arcturus, cosmic guardian of the Great Bear. The round table symbolises the zodiac. King Arthur sent his twelve knights, representing the twelve signs of the zodiac, on a quest to discover the Holy Grail, the Christ Light within. He created Camelot, said to be in south west of England.

King Arthur was given the sacred Excalibur Sword by the Lady of the Lake from the Isle of Avalon. He used it in battle, and died when he allowed his analytical mind to take precedence over his love, instinctive and intuitive nature, i.e., when he gave away his spiritual power.

The Great Bear, part of the old evolution, awakened to the Light and can no longer attempt aggressive, warring control patterns, because the feminine is no longer afraid of his might and weight. She cleansed his cave, filled him with light and love and encouraged him to awaken humans to the light of higher consciousness by activating formerly dormant extra-sensory perception.

Ascended Master Djwhal Khul (D.K.) The Tibetan

I experience D.K. as having tremendous patience. Stable, kind, loving, serene and calm with a great depth of insight and wisdom. Buddha-like with a mysterious characteristic. The main quality he taught me was Wisdom.

Pegasus

Was one of my galactic transport guides. He morphed from Chiron, a centaur with an animal body, into a white Unicorn with wings. The Unicorn is the highest symbol associated with Capricorn. The horn of the Unicorn symbolises an active third eye; clairvoyance.

Without his love and guidance this book could not have been written. He activated my soul's memory and lineage. Always ready to serve the Divine Plan for human evolution, he provides transport into higher dimensional realms to activate soul memories.

Ascended Master Beinso Duono (Peter Duenov)

Through teaching groups his sacred circle dance Paneurythmy for 14 years (half a Saturn cycle) the movements formed geometric patterns replicating those in the heavens. The love and joy created through the groups dancing in unity opened up ancient knowledge and memory portals at the global sacred sites visited.

The main quality he taught me was joy through movement and dance.

Aquila: The Eagle

The eagle was another of my early galactic transport guides. He flew me into the expansiveness of space with my inner eyes wide open. From my higher vantage point on this star, I could see how humans were wandering around like a herd of lost sheep, needing a loving, all-seeing leader of light to unite them as one.

The Eagle is symbolic for the human process of developing an observer and loving perspective to self and life.

Human Ashtara:

Curious, intuitive, adventurous, practical, sensitive, caring, eager and open to learn.

The main quality I exhibited was curiosity - the desire to know, appreciate and understand the bigger picture. There always is one.

Orion: Al Nilam, the central star of Orion's belt.

Our consciousness raising process accelerates when we pass through the central star in the belt of Orion to experience the 'all-seeing eye'. We then have the eyes to see, and to experience, the infinite dimensions beyond.

The Galactic Federation of Light work from Orion. Ascension into higher frequencies of light requires warrior energy of focused intent, spiritual will, self-sovereignty, and desire to attain this higher frequency of Light.

Star Master Omega: spokesperson for the Council of Nine. Calm, authorative, intelligent, considerate, caring, loving, self-assured and compassionate.

Spokesperson of the Council of Nine, Omega always stood at the centre of the open, top section of the horseshoe training space: Extremely tall and thin, with long arms and seemingly over-sized large hands, he appeared to glow from within with a soft silvery coloured light.

Emanating calm authority and sovereignty, he appeared at all times to be attentive, considerate, caring, loving and self-assured. His energy was of a true leader. The main quality I perceived in him was Intelligence, arising from a combination of the Higher Mind and Divine Love.

Pisces Australis - The Great Fish

Symbolises the power of the 'cosmic waters', the Mother Goddess. Jonah, swallowed by the whale and released because of his decision to focus and act upon his internal spiritual guidance, rebirthed into a light body through spiritual conversion in the cosmic ocean of the Goddess. The big fish is the great goddess in disguise, situated between Capricorn and Pisces is the area of space referred to by the ancients as the 'Cosmic Waters'.

The Arcturians told me the akashic records are now stored in the akasha in the etheric field of Pisces Australis situated in the vicinity of Byron Bay NSW, Australia. Byron Bay is the first place in Australia to receive the light of the Sun.

The Sirians gave us the whales and dolphins for their sound frequencies, codes and chords to assist our evolution into higher consciousness. The whales and dolphins pass by the beaches of Byron Bay in their annual journey north and south from and to the Antarctic.

Sirian Healer Athenia:

Her Virgo precision, attention to detail and order are needed to maintain the earth's and each soul's akashic records.

Efficient, effective, precise, organised, she is available to assist the healing of discordant energies that arise within human minds during our ascent into Christ, or galactic consciousness. The main quality she taught me was discernment.

The Southern Cross Constellation

Symbolises the need for we humans to become as a small child, reclaiming our innocence and wonderment, so we can access joy, natural outpourings of creativity, delight in Nature, divine Love and Light. Just like the Rainbow Children of Light, we humans can consciously and with self-awareness, play the game of life through love, wonderment, joy and spontaneous creativity expression.

The Southern Cross constellation of stars was important to the ancient astronomers of the Andes, Maya and the First Australians. Ancient Andean wisdom has the Southern Cross as their most revered night light. A stone image of it was found at Machu Picchu in Peru. The Inca knew it as the Chakana, which means 'the stair'. The stairway to heaven and higher consciousness?

Part of our human evolutionary ascension process is to re-member the parts of ourselves that were cut off and separated in times past. This soul retrieval process enables us to re-claim our true identity. We were each designed as a unified being of light connected to Divine Intelligence through universal love. The human heart holds the pearls of wisdom and is the key to an illumined, fully conscious mind.

Ascended Master Jesus: (Yeshua ben Joseph)

Love, compassion, faith, consideration, calm self-assertiveness, authority and wisdom. During my meeting experiences with the Council of Nine I found it difficult to regard myself as equal to him, even though he kept encouraging me to do so saying *I know you.* I felt I knew him from times past, but it took me a long time to accept his request for equality. The main qualities he taught me were Compassion and Humility.

Many children are now being born far more self-aware than their parents. Sometimes known as Crystal, or

Rainbow children, they are coming from different star systems and galaxies to earth at this pivotal time in human evolution to accelerate the collective human consciousness into greater light.

They have a lightness of being, joy in their hearts and love for the One Source. Conflict and war have no place in their lives. They avoid heavy conditions, foods, systems and also over-bearing people. Masters of reconciliation, they want peace, and know how to create it.

They have an advanced consciousness, often are telepathic, and often use imagery combined with words to relay their messages. They encourage us to go beyond our mental belief tunnels because they envisage a more wonderful future, and will teach us to create it. They have no interest in the past. They come to teach the principle of Oneness. Separation has no place in their lives.

Yeshua has a way with children, knows who the new children are and why they've come. He guides them into full creative expression. As they become adults, they embrace the human ascension process through love, compassion, faith, consideration and wisdom. They are connected to, and can become, the Rainbow Warriors, embodying the Rainbow Colour Rays of Creation.

Operating through conceptual thought and intuition, they will create new education systems to aid human development of consciousness. They are incarnating in increasing numbers and are our future.

*

The Arcturians say:

In times to come, when the sleeping ones awake, a new world will emerge, created by the collective mind. A world of joy and wonderment for the grandeur and scope of the Prime Creator's Plan for the evolution of the human soul. This may seem like a children's fairy tale, but it is not.

So many are awakening now to the reality of creation and of how each one creates their own reality by their thoughts. Mind management is the next step and then comes conscious creation.

When sufficient numbers of people create loving kindness so it will be experienced by the collective. This is how a new Camelot, an era of peace, joy and happiness will occur. It is in the making, arising from pure and loving hearts and minds. Beyond the psychological process, where shadows cover a loving heart and soul, is the state of being joy.

Consciously connecting to the Christ Light within, and allowing joy to guide each new forward step may seem out of reach and unattainable. We assure you it is not. You can make this choice today. When you do, realise that all that blocks joy will emerge.

Welcome it because by embracing and acknowledging that little child part of yourself, rather than criticising, it will soon pass. Love that part of you to death. By doing so, your internal channels clear allowing emergence of your soul light. Love is the pathway to the new world.

Never can you be disconnected from your Creator. You may think you are. You have the power to change your thoughts. Will you use it?

*

Space Junk

We humans chose the path we have taken: that of immersing ourselves in the third dimensional plane of duality. Other souls didn't. They remain consciously inhabiting the Universal Presence as Spirit Beings, Angels, Archangels, Star Beings etc. They remained in higher dimensional realms beyond material illusion, continually bathed in eternal fields of light.

We asked them to awaken us into the remembrance of our soul's heritage when it was the cosmic timing to do so. This timing connected to humanity's first physical step beyond our earthly home onto another, the Moon. This great historical event that advanced human consciousness beyond our earthly home took place on July 20th 1969.

At that time, Uranus, the Bringer of Light, was 0 degrees 37 minutes Libra; Jupiter, the planet that brings expansion and evolutionary growth was also in Libra, conjunct Uranus at 0 degrees 37min. These two planets when conjunct, take us further into higher dimensional fields of experience should we consciously allow the process.

Saturn, the planetary Time Keeper and Lord of Karma, was 8 degrees, Taurus 3mins. Both Taurus and Libra are ruled by Venus, the planet of love.

The interesting thing is that, at the same time transiting Chiron was in Aries, the sign it is again traversing at the time of writing, fifty-two years later. A Chiron return to the Moon landing event occurred in 2021. Chiron holds the key to open the locked doors of our human psyche enabling emergence of higher, more advanced consciousness. There's a lot of meaning to be unveiled in this comment. As D.K would say: "Ponder on this".

The Arcturians say:

In all walks of life change is to come. Everything will be affected. How you, individually and collectively, handle the change is up to you. A calm and centred approach will ensure a positive outcome.

What will create this change you may well ask. Cosmic design is the answer.

The planets orbit around the Sun in an orderly and calculable design. All Nature is the same. Humans cause change by their thoughts and behaviour. When collective thought is against Nature, natural design problems arise. It has always been.

The space around Earth is now littered with junk. Failed human attempts to launch space craft creates this junk. The atmosphere around Earth is no longer clean.

What will be the natural result?

Human thought based on greed created the space debris. The desire for power and control turned into greed, to the detriment of all living creatures.

Wake up humans, to what you are creating in the now that will affect your future.

As you read and integrate the above message, it would be wise to take into consideration that every word and thought contains a vibration. A negative thought of lack, resentment, jealousy, criticism, judgement, arrogance, etc. carries a low, heavy and dense vibration.

This vibration travels to locations both inside and outside our bodies. Positive thoughts such as joy, love, delight or wonderment, contain a high vibration and they also travel. When human collective thoughts are dense and heavy, our planet experiences it.

NASA, aware of the growing amount of space debris, created a Pollution Orbital Debris Program Office. To what effect?

When we fully embody the fact that within each cell of our body is the living light of Prime Creator's substance, pure love, we will know peace. When we realise there is a massive army of celestial and extra-terrestrial Light beings working to awaken us individually and collectively into acceptance of this fact, we can invite them to assist us on our journey to higher consciousness. When we know and accept, without a shadow of doubt, we are loved beyond measure, we will feel peace.

Through our combined love and group awareness, we can bring stability to not only planet Earth but also to our Solar System and beyond.

With passion and love in my heart and soul to do whatever I could to bring about peace within me, and within those I teach, I have written this trilogy. Guided by my intuition and use of all my senses, I share my story, as I was asked to do.

The plethora of inner guidance I have had has been extraordinary, and for which my gratitude knows no bounds. I honour my readers and students because you too desire to create peace within and upon Earth, through loving kindness and goodwill. May you focus on your peace journey, consciously realising and releasing all discord, and with dedicated intent, invite Spirit into your lives to assist your journey.

Chiron doesn't enter Taurus until May 2027. There's lots of accelerated global and individual human growth into higher consciousness to take place before Venus, the planet of love, and planetary ruler of Taurus can guide the collective human minds and hearts. It will happen.

How do individuals learn and grow into greater self-awareness? By unconsciously creating emotional dramas and chaos. How does the collective learn? The same way.

The Arcturians say:

And it came to pass that, as humans develop the ability to choose love over fear, and live their lives joyously with child-like wonderment, a new way of living life emerged. The higher vibrations of love attract love. It has always been so.

*

Merlin's Teaching on Perception

Years ago, the Spiritual Master Merlin provided me with a teaching about perception. It too was in a Boardroom. I'm including it here because it has had a big impact on my students' minds when they understand and have been able to integrate the exercise. It can be a relationship saver. It also serves as an introduction to the next part of the Arcturus Astrology de-coding.

Merlin took me into the starry realms and to a round table. Maybe the same round table as that of Arcturus? He demonstrated six ways of perceiving the one thing. The one thing I was asked to view was a pyramid. Merlin asked me to view it from six different angles and to telepathically report on what I saw. I did as he asked.

He suggested that, when about to make a serious decision, it would be wise to view the situation from six different perspectives before proceeding. I recommend you take his advice. It will prevent a lot of anxiety.

*

Everyone sees something different according to personal perception. Other dimensions are available to be seen, yet perception clouds this. A baby is born able to perceive many different dimensions simultaneously. They see light and colours. When a new born baby looks into another's eyes and sees and feels a soul connection, the baby is recognising the other's soul light and energetic emanation.

It takes some time for the baby to close down from its wider perceptive vision to see only the 3D world. They perceive their external world via colour, light, vibration and feeling, soul light and energetic emanation. i.e., they have extra-sensory perception.

Infants also have inner Sight (clairvoyance), inner hearing (clairaudience and telepathy) and inner and outer feeling as body sensations (clairsentience). Apparently, these natural gifts and talents close down as the child identifies with language.

Water is a carrier of consciousness. The human body is comprised of approximately 80% water. Human blood contains water. Water is the element governing emotion. Emotions are stored in memory. Through the mechanism of the heart, consciousness flows through the blood and is passed into other organs, as if in sound waves. All information is in the form of experience.

The emotions felt during these experiences are transported through the body via the blood vessels from the heart. There are many miles of these channels within the human body. Psychological density blocks the blood flow.

The hippocampus is connected to the heart and is the part of the brain that stores and awakens memory. The more the heart is filled with the feeling of love, the greater is the recall of memory. This memory is not only of this lifetime but of other experiences in different places, dimensions and times.

The embodied soul begins to remember its past history. When all bodily organs are fed with the consciousness of love, they too function at their highest capacity.

The energetic field of the heart can be measured. This field can extend for at least a metre out from the body. When the cells within the heart entrain, this field expands to even greater distances. When two fields entrain through feelings of love, a huge single field is created that can be felt and seen by those who have the eyes to do so.

Now let's connect the senses mentioned above to the organs of perception relative to the Arcturus Astrology system:

Sensory and Extra-Sensory Perception (ESP)

We in the higher realms have a different perspective from humans. Because we view from above, we see holistically. It is well that our scribe has been trained to also have this perspective. Each person views life through their own lens. This lens is shaped by experience. Societal and parental conditioning result in a narrow perspective.

Perspectives can change. We endeavour to bring about positive change by providing you with a broader view to life.

Many people, unwilling to change their perspectives, suffer stagnation that leads to health problems. The Universal Law of Rhythm and Movement is always in effect. Thoughts travel as does every emotion.

Planets are continually in motion around a central star. Storms develop through atmospheric movement. Universes are continually expanding. Humans are designed to move, not only physically but also mentally and spiritually. Stagnation is alien to the Universal Law of Rhythm and Movement. When it occurs, problems follow.

The Arcturians speak:

"We are always with you, observing, recording and sharing our energy, our light and love. You can call on us to assist

with personal matters if you wish. We are here to serve Prime Creator's Plan.

We are a living consciousness without form having evolved beyond the need for a physical body. In times to come you will do the same.

Our message today is one of love. You were created through love. It is your essence.

Love is a felt sense, connected to your soul. You feel the essence of love in your heart, an organ of perception.

At this phase of your evolutionary journey to the Light of higher consciousness, you will more readily feel the love essence connecting to your soul. It will feel joyful. To our scribe, this feeling is so joyful it almost brings her to tears. Occurring during times of focused relaxed creative expression, the joy is contagious. Others also experience the shared joy. When you live lives following the path of joy, and love of your creative expression, your body responds with good health.

We urge you to find and follow your path of joy.

We are the Arcturians. We love you dearly and will come again.

They continue:

In times to come, humans will realise their connection to the stars and will view them differently to now. Multi-dimensional viewing will become normal. Currently, babies are born with multi-dimensional perception, able to see colours and light beyond the normal range. Losing this ability within their early years, the growing child adapts to its environment.

As more humans become self-aware, their sensory perception changes. Able to see and feel life beyond the normal range becomes natural. Rather than relying on their analytical ability to make sense of their reality, higher sensory perception will take precedence. This is Nature's way. Re-adapting to embrace Nature and to live according to natural design is the way of your future. Nature is the teacher. Perfect and orderly is Prime Creator's design.

Humans were designed with sensory perception to live life in harmony and peace attuned to Nature's design. Through the gift of free will you decided to follow the dictates of the analytical brain. Indigenous races maintained their connection to Nature.

Nature heals. Nature provides food, shelter and medicines.

A return to Nature is imminent. When humans appreciate the value of Nature's beauty and bounty, Nature will respond in kind.

We love you dearly and will come again.

*

Through fear, I took a long time to de-code the Arcturus Astrology System as I had been asked to do. Years went by. And then I realised that esoteric science connects to, and combines with, ancient Sanskrit philosophy. Having been a student of yoga and still regularly attending two classes a week, I've learned to place an extremely high value on this ancient teaching, it's philosophy and associated practices.

The philosophy behind yoga is geared to raise human consciousness by clearing psychological shadows from the body and mind. Unity of body, mind and spirit is its aim. Dense psychology stores in different parts of the body. Sixty-four years ago, I began practicing yoga postures to aid natural childbirth. It worked. Three babies born naturally. No hassles or struggle. Fear didn't enter my mind. This is the key.

According to ancient Eastern philosophy, there are two great energy centres in the head. The centres D.K. refers to are the chakras. One of them is situated between the eyebrows. (Ajna or Third Eye chakra). The energy in this chakra blends with that of the five chakras below it.

The other in the head centre (Crown chakra) is awakened through regular meditation, service and aspiration. It is through this centre the soul can connect with the personality. This head centre connects to spirit, a positive (like a battery) energy charge masculine aspect, just as the centre between the eyebrows is a symbol of matter, the feminine negative energy charge. Connected with these force vortices are two physical plane organs, the pituitary and the pineal gland. The pituitary is positive, the pineal is negative. These two organs are the higher correspondences of the male and female organs of physical reproduction.

Our senses receive continual communications in the form of informational energy. They are living organs.

An Oxford dictionary's definition of an organ:

a) *A musical instrument having a special tone*
b) *Part of animal or vegetable body adapted for special vital function e.g., organs of speech, perception, digestion, generation etc.*

The heart is the main organ of perception in the human body. It sees and feels. It knows how and when to distribute information to other parts of the body. The heart is intelligent. It perceives at a rate much faster than the human brain. The feelings emanating from the heart can be picked up and amplified by another being.

When two people connect with each other through heart-felt love, entrainment, or synthesis, occurs. The two hearts pass encoded information to the many other body organs, flooding them with the energy entrained. The human heart has an innate intelligence. It connects to other parts of the body via its web of inter-connecting channels or nadis. It is the central hub of a system of management and is self-regulating.

Indigenous people of all cultures understood heart technology. They lived and experienced the natural world by tuning in to their senses. Modern civilisations disregarded this natural intelligence.

Awareness is focused through the senses. By attentively observing ourselves sensing and feeling, we reconnect to the natural rhythms of Nature. Our five senses can be thought of as primitive receptors. They only pick up a small amount of data of the world around us.

Our extra-sensory perception is finely tuned to the natural world. By consciously using our finer senses many of the mysteries of the universe can be unravelled.

You can find all wisdom you will ever need by using your extra-sensory perception and intuition to uncover from within your own mysteries. Direct revelation can be yours should you be willing to practice developing extra-sensory perception.

Internal peace and harmony can result when you release old beliefs that create stagnation. Our natural senses are our survival guide and are synonymous with our Spirit's presence, occupying the same space.

In the new world that awaits, we will need to utilise our atrophied natural ESP gifts because they could be life savers. Learning to trust them is imperative. Many of the young children being born now are using them, and teaching their parents to also do so.

*

According to Esoteric Psychology Vol 1, The brain is responsive to the seven senses as:

1. Hearing
2. Touch
3. Sight
4. Taste
5. Smell
6. The mind, the common sense
7. The intuition, or the synthetic sense.

Through these seven senses contact with the world of matter and of spirit becomes possible. *The seven senses are, in a peculiar way, the physical plane correspondences of the seven rays, and are closely related and governed by them all.* (D.K)

The brain can be thought of as the eye of the soul, looking out onto the physical world.

My task is to connect the senses and organs of perception to Arcturus Astrology. Ten cosmic bodies. Ten chakras, ten senses, ten organs of perception.

Organs of Perception and Chakras

1. The ears are an Organ of perception, and of Extra-sensory perception: Hearing: Clairaudience, telepathy, and the voice of conscience. (Throat chakra)

2. The skin and spine are organs of perception. Touch affects the skin, as does truth when intuitively spoken or revealed i.e., goose bumps over the spine and body. Clairsentience is the response to vibration. The awareness and recognition of that which lies within and without the physical human body.

3. The eyes are Organ of Perception: Extra-sensory perception: Clairvoyance. (Third Eye chakra).

4. The tongue is an Organ of Perception: Taste. (Throat chakra)

5. The nose is another Organ of Perception: Olfaction, the sense of smell. Like animals, we human animals unconsciously smell others, and connect or otherwise with another through this perceptive sense. This subtle process is closely connected to the way we experience new relationships. (Throat chakra).

6. The Mind is an organ of perception. D.K divides the mind into two categories as: The lower and higher mind. The lower mind is that of intellect, logic, rationality and societal conditioning.

7. The Higher Mind: The common sense is the "seat of consciousness, thought, volition and feeling". The intuition or the synthetic sense is an organ of perception. (Synthesis: combination, putting together.) It is the higher mind that connects to Sirius and our extra-sensory perception. During the process of human conception, God seeds were planted into our Higher Minds and our hearts. (Third Eye and Crown chakras) Organs of Perception: Pineal and Pituitary glands, and the Hippocampus.

8. The Navel. Feeling Perception (a): Gastro-intestinal tract (G.I). Where emotions such as fear and anxiety are often experienced. (Solar Plexus chakra). I often wonder about the psychological effect on a new-born baby when the umbilical cord is cut. It symbolises a disconnection from the world of Spirit into the world of form. Stress must be involved.

9. Heart: Feeling perception (b): Scientists at the Heart-Math Institute in California connected to the Stanford University made a discovery that the heart has its own brain, with brain cells. This brain is small only about 40,000 cells, but this discovery is one that indigenous people and ancient wisdom holders, have always known. The heart has its own brain, and the heart generates the largest and most powerful electromagnetic energy field of any organ in the human body. The scientist found that a toroidal field (shaped like a donut, a torus) extends for about eight to ten feet in diameter from the heart's axis. (Heart chakra)

10. Integration: Mind, Body and Spirit: Heart/higher mind connection with Great Spirit. (Heart and head chakras)

The first senses developed in a child are hearing, touch, and sight. These are the three vital senses. Taste and smell follow later.

On the path of inner development, the sequence is the same. Hearing responds to the voice of conscience, as it guides and directs. This covers the period of normal evolution. An evolved soul hears the inner voice of the Higher Self, and of Spiritual Masters and other Light filled beings.

Our bodies can be symbolically likened to an orchestra, with each of the Organs of Perception a different musical instrument. The Soul is the conductor. The higher mind, the seat of consciousness present in every cell of the body, contains the knowledge of our soul's experiences.

When all the different sensory instruments (organs) are 'in tune', our Soul resonates with the self-created harmonics. i.e., the Soul responds to the personality's love-based choices. These love-based choices vibrate at a higher frequency. When each organ of perception is operating at optimum efficiency, we hear the inner 'Music of the Spheres' and the body responds with vibrant health.

We each perceive our world through our senses. We, as a living organism, receive and broadcast electromagnetic signals, many of which contain a great deal of information. Radio waves carry information in much the same way.

We learn to decipher the information through our senses, according to our level of psychological functioning. All living organisms do the same. Our cells recognise the differences in these signals, and decode them according to our vibratory speed, our level of consciousness.

Our brain cannot compute sensory perception. It functions as a computer. Our thoughts feed it information. The information we feed it consists of linear conditioned beliefs based upon societal and parental programmes. The brain does not have a moral code, a conscience.

The brain is an important organ because it processes data that connects to the central nervous system. Data in and data out. The thoughts we think affect our central nervous system and our psychological and physical health.

The heart is an organ of perception, a most powerful electro-magnetic receiver and generator. It feels truth. The brain cannot feel.

The human evolutionary revolution, the shift in human consciousness expansion, is based on sensing and feeling, rather than thinking. It's time to come to our senses, to use common sense and our organs of perception to guide our lives. This is what we were designed to do. Spending quiet time in Nature assists this process.

As we reclaim our ability to feel and sense with our heart, the psychological barriers built up over many life times, like scar tissue around the heart, gradually dissolve.

Love is the foundation of divine intelligence. The brain, like a computer, receives sensory input from the body, mind and spirit, computes it and produces a result. It cannot feel the energy behind loving thoughts, the heart/brain connection.

The heart is usually the first point of information impact. It feels emotionally whatever we perceive. Our experience of the world first comes through the heart which feels and considers the experience. It then sends data to the brain for further processing. The brain computes that data according to its pre-programming.

The heart then receives the return brain information and decides on the action to take. It can feel into the old programming, or choose to act according to its initial heart-felt experience. Between the heart/brain connection, a decision is made.

The body and mind are interlinked and interconnected. The brain is a part of the body. The mind is not the brain, but we can open our minds by using our brain. How?

Breath work is one way. The brain needs oxygen to ensure it works well. In meditation, we can focus on sending deep and long in-breaths, filled with the Creator's love, to the brain. Eventually, this raises our consciousness through brain expansion, which can be felt as a sensation. It is a psychological process.

According to author Stephen Harrod Buhner in his enlightening book, *The Secret Teachings of Plants – The Intelligence of the Heart in the Direct Perception of Nature,* the heart is one of the major endocrine glands in the body.

He also says that the hippocampus is an organ in the brain sensitive to the fluctuations in magnetic fields. Why? Because it contains magnetite, therefore is an attractor. Involved in interpreting memory, including our soul's memory which stores sensory experiences, it is able to extract meaning from these experiences, so is closely attuned to the health of our heart.

Our senses are living organs. They receive continual communications in the form of informational energy. Indigenous people of all cultures lived and experienced the natural world by attuning to their senses. Modern civilisations discarded this natural intelligence.

As sensing takes precedence over thinking, the body becomes more alive. Awareness is focused through the senses. By attentively observing all that is sensed, we begin to attune to the natural world, as did our ancient ancestors.

Australian Aborigine wisdom holders say that our Earth now moves on different Dreaming tracks and the whole universe is moving and changing. Nothing stays still. These Dreaming tracks are not only on Earth but also in the Solar System and beyond.

They say that the Dreaming tracks are all about self-knowledge and self-consciousness, and that the Dreaming tracks carry the essence of the Creator's divine energy. Because we humans also carry the Creator's essence, we have the potential to reveal divine characteristics.

The Aborigines say our ancient knowing sense has been totally dishonoured. Head logic and computers have taken its place. And, that our energy field is centred just above our navel, (solar plexus chakra) where we feel negative emotions such as betrayal, greed, shame, rejection, abuse etc. These dense feelings reach all the way to our soul/spirit. Silt and pollution build up in our energy body via negative thoughts and feelings.

Over the last few centuries, the Earth's positive Dreaming tracks began to lose energy and close down because of human destruction created by dark thoughts and intent. We have been in a spiritual coma, feeding our egos and declining into spiritual lostness. The spiritual elders did what they could do in 1990, guided by their inner knowing, senses and feelings.

To correct the situation, not only in Australia but all around the world, they broke 300 old and special stones opening an incredible vortex to enable the descent of greater spiritual Light. Many people began to wake-up all around the globe.

Prior to the breaking of the stones, Australian aboriginal female wisdom holder, Minmia, was given a spiritual assignment. She was to travel around the world to specific areas to find and re-claim some of the unique stones that needed to be broken. To accomplish this spiritual mission, she needed to feel and sense her way, using her gut instinct and intuition. The deed was done, following a circuitous route. She needed to quash her logical brain and its negative thinking to accomplish her spiritual planetary work. ref. *Under the Quandong Tree*, Minmia.

Now to some scientific discoveries that connect to my experiences and helped me to make sense of my reality, and maybe yours.

In my book, *I Am an Experiment*, I mentioned that NASA scientists test humans to see who best can be used for their space agenda purposes. In Drunvalo Melchizedek's book - *Living in the Heart,* he shares the story of a completely blind woman, who technically has no eyes, yet she sees life clearly and is able to manage it well. NASA scientists heard about her and asked to see her for testing purposes.

Their first test was to ask her what she was seeing inside her head. She told them she was moving through space and continuously watching what was going on in the solar system, saying she was restricted from seeing beyond it. They asked her to move alongside one of their space satellites to give a reading of what she saw. She did it precisely. They employed her.

The way she sees is through an inner image similar to a TV screen. Situated on her inner TV screen are small, framed images placed around the outside of the main large central rectangular screen. The images inside the small screens are rapidly moving. Some have extra-terrestrial faces, and some have wavy static frequency lines.

What is being described here is similar to one of my meditation experiences, I wrote about in *I Am an Experiment* when I was taken aboard a Sirian spaceship. The first thing I observed, after being warmly welcomed by a gracious female Sirian, was a group of tall, slim and slightly blue Sirian Light Beings working on one large white horseshoe shaped desk, each with their own computer screen. I did not see any cords, cables or power points. The entire curved desk was clean and cleared, except for the computers. What I saw on their computer screens was similar to the description above.

Drunvalo also writes about the super-psychic children who attend special training schools in China and Russia. The governments of those countries apparently fund these schools because they place a high value on the children's extra-sensory perceptive skills.

In a monastery, high in the Bulgarian mountains, the resident monks teach the children to see with their inner screen, and with the different parts of their bodies. Their primary message is that peace lives within each of us.

*

The Perceptive Chakra System

The Arcturians say:

When one ascertains the truth of their soul's longing, a process begins to fulfil that truth. To ascertain this truth can take much in-depth work, and many visits into Pluto's subconscious domain. The truth of the soul's longing is contained within a treasure chest hidden in the depths of the subconscious realm.

When an individual is ready to begin their soul's work in the world of man, the pearls of wisdom, gathered over numerous life times, are released from this chest and float to the surface of consciousness. The pearls of wisdom are then available to be utilised in the soul's work. It is so.

Within the human body are many meridians, or nadis, that are vehicles of expression for various energies and intelligences. Intelligences from other dimensions move in and out of these nadis in an endeavour to mine information.

These channels contain information of experiences that took place over thousands of years and this information is valuable. Intelligences from other dimensions use this information for their data base and store it in libraries. They do not have these experiences, so choose to learn about them through riding the light waves within the human body.

*

Chakra is a Sanskrit term meaning wheel. A chakra is a pulsating, spinning vortex of life force energy that has a regulating effect on specific organs and glands as well as on our emotional and mental bodies. There are seven of these spinning wheels operating within our etheric body.

All chakra centres in the human energy field are the repositories of various forms of energy including sensory. They operate as perpetually rotating wheels into which life-streams enter. All are connected. The chakras function as energy transformers. They receive, absorb and transmit life force energy stepped-down from higher Light frequencies.

Just as our planets act as transformers or power stations, stepping-down their energy to planet Earth and all living species, so too do the chakras step-down cosmic energy in the form of information and consciousness to various parts of our bodies.

Universal life-force energy is received and distributed through each chakra. Each of the frequency levels manifest as a colour. The colours associated with each of the higher chakra centres vibrate at a higher frequency to that of the lower chakras.

For example, the third eye chakra, the colour of violet, vibrates at a frequency twice the speed of the base chakra, which is red. Each chakra spins at a different frequency dependent upon the particular qualities of consciousness.

When each chakra spins at its natural and perfect frequency, as it is designed to do, the body radiates perfect health. Emotions are harmonised and we enjoy a deep sense of inner peace and calm.

The opposite is also true. If we lead unhealthy lifestyles separated from Nature, are unconscious and lacking self-awareness, we become emotionally, mentally and physically stressed and our chakras become blocked. Ill health and disease result.

The chakras function to maintain the body as a vehicle for light, love and higher consciousness. Invisible, unable to be seen by the naked eye, they are even more subtle than are atoms and particles. An invisible atom contains a tremendous amount of energy. Its nuclear effect, when turned into a bomb, has been seen, photographed and experienced. It is a fact that cannot be denied.

Ancient eastern yoga wisdom views the body as dual with a gross material body and a subtle energy body. According to Shyam Sunda Goswami in his book, *Layayoga, The Definitive Guide to the Chakras and Kundalini,* the material body is composed of flesh, bone, hair, blood etc. and undergoes metabolism.

The subtle energy body, operating like an envelope around the physical body, is composed of nadis, which are force-motion-lines. Ida is the nadi that is silvery white, like moon light; is feminine in nature, and situated on the left side of the energy body. Pingala is like the sun, masculine. Between these two nadis is Sushumna energy, containing Brahma-nadi. (Brahma is the Sanskrit name for Divine Power).

Sushumna is extremely fine energy and is situated outside and in front of the vertebral column. The chakras are situated within the Sushumna nadi. Inside this nadi is the extremely subtle chitrini nadi, divine in character and is in the form of letters. Hmm! This takes me back to my introduction, and author/scientist Greg Bradon's findings of the God Code within every human as letters. It's interesting to contemplate that the ancient Hebrew language is one of divine fiery letters, and is similar in theme to the Sanskrit knowledge wisdom system with its representation of letters.

Where did this ancient knowledge/wisdom system come from? My view, Atlantis. Those who carried the sacred wisdom escaped from Atlantis before their human counterparts destroyed it, travelling to different countries to maintain and share it. Some went to Asia, some to South America, and some to Egypt. One stream of knowledge/wisdom and at least two ways of expressing it. On the other hand, the Australian Aborigines are the only remaining race of Lemurians, a race that existed before Atlanteans.

Also, consider this. Why did Omega show himself to me as silvery, moon-like energy and RA as sun-like energy? Is the system I'm working on bringing through a combination of both Egyptian and Sanskrit teachings? East and West connecting as one?

Let's take this a bit further. The material body is the effect of the metamorphosis of the basic energy of prana-force. Prana is matter free and exhibits a circular wave motion reducible to a subtle infinitesimal point. Different forms of energy active in the physical body are entirely dependent on the operation of the pranic forces.

Energy in a more refined form appears as electrical. Consider it is this internal electrical energy that activates truth 'goose bumps', and all the 'sparks' of light I've mentioned throughout this trilogy.

Cosmic light can only enter the human body via the chakra centres.

Therefore, the Cosmic Rays enter the human chakra centres, activating them into awareness.

*

Each time I asked the Council of Nine to assist me in uncovering the intricacies of the Arcturus System, they gave me the same experience. I felt tremendous light sensitivity, heart-felt love and harmony from their love. I could feel, through body sensations, where Light energy flowed into my crown chakra, moving instantaneously to my heart. The love energy I felt in my heart travelled to the solar plexus chakra. The other chakras became energised during the transduction process.

The soles of our feet are also energy centres, and buzz when higher vibratory energy is introduced, as do the other chakras. Our feet connect us to Mother Earth. I often feel Gaia's vibrations through my feet, even when wearing shoes.

Chakra One is the base, and includes our feet.
This chakra is the foundation, or prime electrical ground for the body. It connects the body's electro-magnetic grid to earth. Situated in this chakra is a square, symbolising earth. Within the square is a triangle connecting the fire of our Spirit, our psychology and consciousness.

Chakra Two is the Sacral Plexus.
The sexual or creative centre, located in the same area as the genitals. The two lower chakras contain the kundalini (serpent) energy, the vitalising energy force that invigorates the body when activated through soul memory, self-realisations and self-responsibility. Referred to as the Rainbow Serpent by Australian Aborigines, she uncoils from the base of the spine to emerge through the crown chakra in blissful ecstasy, unifying with her heavenly twin soul. In perfect harmony, they dance in a circle as the ouroboros' in a circle of cosmic light.

Chakra Three is the Solar Plexus,
Acts as the body's lower emotional centre, located in the region of the naval, spleen and liver. This chakra contains the silver cord (like an etheric umbilical cord) which connects to our soul force. The silver cord contains the vibratory electronic patterning of each soul. This chakra is also a regulator of kundalini energy.

Chakra Four is the Heart
This chakra is the where the higher emotions such as love, joy, child-like innocence, and wonderment are experienced. It is centred around the heart and lungs.

Chakra Five is the Throat
The communication centre located in the region of the larynx. Through feelings of love that arise in the heart while regularly toning sacred sounds, we create healing frequencies. The loving sounds created travel internally to discordant energies in our bodies, enabling healing. Through regular love-based toning, we may suddenly begin speaking or toning our soul's language, as a natural response to love and joy.

In my workshops, I take students through the simple process of generating loving feelings in their hearts, and then ask them to tone vowel sounds from the love vibration created, and to concentrate on them. I ask them to relax into the vibratory stillness created when the practice completes. They experience bliss.

Many do not want to come out of the experience. The vibrations created tune up the energy field and penetrate into the physical body.

When toning our own sounds from loving feelings in our heart, our body has a greater opportunity to heal discordant energies. And, we also purify the air around us through the invisible energetic geometries formed from the combined sounds. These vibrational sounds travel in the body to wherever they are most needed.

Chakra Six is the Third eye
Located above the nose in the centre of the brow. It is the centre for receiving and utilising visionary and psychic experiences. When opened sufficiently, it can receive higher light frequencies. Physicist, J.J. Hurtak in his book,

The Keys of Enoch, The Book of Knowledge, says that the third eye chakra is associated with the pineal gland and can be considered a rudimentary eye.

He says that the awakening of this chakra constitutes the beginning of the spiritual journey to oneness and the beginning of cosmic consciousness. As it develops, it joins with the chakras directly above it like a stem and flower. I like to imagine we are stem cells, creating new cells according to our higher levels of consciousness.

Hurtak also says that through the pineal area is a third eye network. This network enables the Brotherhood of Light to project mental signals with Light geometries so they can be picked up by our minds' perceptual apparatus.

I can vouch for this because I regularly receive strange colourful images, seemingly framed with bright light behind them. I know they are consciousness accelerators and respect and honour their presence in my third eye. Light is the foundation through which, and by which, all higher forms of energy are transduced.

Chakra Seven is the Crown
Associated with connection to our Higher Self, it is where the physical body can connect to spiritual energy and begin to open up the higher mind and lighter vibratory states of being.

Chakra Eight is above the Crown
I refer to it as my platform, experiencing it as approximately a metre above my crown. According to Hurtak, it is the template through which the higher

energy systems in the body's field unite with the human biocomputer system. This system is controlled by the seven chakra stations. The eighth chakra allows for the unification of the Overself Body and the human biological system. Through this chakra we are able to experience Divine Light appearing over our heads.

Chakra Nine is the Galactic Chakra
From it we open our consciousness to experience multi-dimensionality.

Chakra Ten is the Cosmic Chakra
The amalgamation of higher and lighter energies from which the transduction process occurs.

We are each in the process of learning to release the dense psychology associated with each of the chakras developed over our soul's many life-times.

The heart chakra, with its main debilitating theme of separation from Divine love, appears to be the greatest challenge. The spiritual growth process of purification through self-realisation of weakening cause, and the taking of responsibility for creating the weakening cause, enables activation of the higher chakra centres.

The memory centre is associated with the retrieval and reclaiming of our soul's ancient wisdom and knowledge, through remembrance of it. For astrological Ages of time, this part of the head centre atrophied.

With the new light energies bombarding our planet, and through regular spiritual practices, we can restore it. We can experience vision-like states of being during our meditations and can learn to interpret and intuit their symbology via this centre. We can develop the ability and the trust to retrieve our soul's akashic records. There are many alternate health practitioners now doing this soul retrieval work.

The three additional chakras in this system - eight, nine and ten - begin to function when:

- 1: We consciously, with dedication and commitment, open our hearts and minds to developing self-love, self-value, and self-awareness

- 2: Have manifested love sufficient to become conscious of the essence of Divine Creator within, and desire to serve this Source of Great Light and our fellow humans.

- 3: Are willingly purifying our depth psychology,

- 4. Have learned to concentrate so deeply we are able to quieten all mental chatter.

- 5: Willingly take personal responsibility for our creations.

This spiritual evolutionary process triggers the physical body into loving vibrations. This is needed before the body can receive the higher frequencies of Light. This means that all the seven chakras must first work in perfect harmony before alignment with the eighth chakra takes place.

The higher and finer chakras are aligned with bright, clear electric type colours representing the energies of higher consciousness. They are difficult to describe, and I did my best to do so in my two previous books. The chakras in and above the head connect to our interdimensional capabilities.

This connection begins in the chakra between the eyebrows, the third eye chakra. In this energy centre, operating as a junction, a change takes place in the distribution of three lines of pranic force. In other words, the Ida and Pingala meet and cross over the Sushumna nadi, connecting as one.

Reflect on the Caduceus, the medical industry's symbol of two energy serpents, each side of the spine, Ida (feminine) and Pingala (masculine) rising from the base chakra and crossing over at the heart and third eye chakras before they reach the crown.

Why then does modern western medicine disregard the Eastern Sanskrit wisdom teachings? Quantum physics teaches us there is no difference between energy and matter. e.g., the physical and the energy (etheric) body. All systems in the human body, from the atomic to the molecular level, are constantly in motion attempting to create resonance.

This resonance can be felt as vibratory. Tuning into it expands our understanding of how subtle energy directs and maintains health and vitality in the human body. Alternate health practitioners base their healing practices on energy medicine. When will allopathic medicine choose to merge with physics, biology and biochemistry to embrace understanding of the human energy field?

An added energy system of geometric force emerges when human consciousness is ready to receive it. Within this expanded chakra system has been assigned various Deities for each chakra.

Sanskrit Deities – hmm? And I'm to assign each member of the Council of Nine to a chakra? Whatever. I'll do my best.

The energy centres from the throat below the Ajna centre are designed to aid the bringing into the body complete operational efficiency. That can only occur when psychological density is realised and consciously alchemised.

*

The Chakras, Council of Nine and the Star Satellites

Our chakras are organs of perception.

Tenth Chakra: Sirius, the central operating system, receives comic light Rays from Arcturus.
The unified Council of Nine oversee the distribution of these Rays to the nine satellites as:

Ninth Chakra: Omega
Oversees the transmission of refined and loving consciousness energies to Orion:

Eighth Chakra: RA
Oversees the transmission of these Rays to the Pleiades.

Seventh Chakra: Crown: (Sahasrara). Ascended Master Sai Baba
Oversees the transmission of the Rays to Antares.

Sixth Chakra: Third Eye: (Ajna) The Tibetan, Djwhal Khul (D.K.)
Oversees the transmission of the cosmic Rays to the Great Bear.

5th Chakra: Throat: (Vishudha) Ascended Master Beinso Duenov (Peter Duenov).
Oversees the transmission of the cosmic Rays to Pegasus.

4th Chakra: Heart: (Anahata) Ascended Master Jesus:
Oversees the transmission of these Rays to The Southern Cross.

3rd Chakra: Solar Plexus: (Manipura) Sirian Light Being, Athenia:
Oversees the transmission of these Rays to Pisces Australis, the great Fish.

2nd Chakra: Sacral: (Svadhistthana): Feminine Buddha, Tara.
Oversees the transmission of these Rays to Draco, the Dragon.

1st Chakra: Base or Root (Muladhara): Ashtara
Oversees the transmission of these Rays to Aquilla, the Eagle.

*

The human chakras are organs of perception. As they receive cosmic Light, they ignite the physical organs within them into activity. As we awaken our higher chakras, from heart upwards, our extra sensory perception also awakens. It takes dedication and commitment to maintain this finer cosmic Light.

The chakras, like etheric disks, are anchored in the glands of the endocrine system and these glands bridge spirit and matter. They are biological receptors that translate the higher frequency information collected by the chakras into a language the nervous system can understand. They are damaged by violent emotions such as hatred, resentment and rage.

When our love is withheld, it must be returned again to our God essence to regenerate. Our Light and Divine Love influences all within its sphere of influence, not only in this third dimensional realm.

When we release our heart to love, an etheric antenna unfolds in our light body that surrounds the physical body like an envelope. It behaves like a sensitive plant, providing us with an etheric organ of perception.

The fully activated human sensory system is a cosmic structural pattern replicated in planetary, solar, galactic and universal fields. For those humans awake and fully aware of the Divine force within them, this sensory system is capable of bringing through its own circuitry system the awareness of the angelic and star beings who over-light each of these fields.

The Prime Creator's love is the glue-like substance that holds all of life together. By awakening to divine love, we humans generate the ability to access universally compatible systems. We open ourselves up to become cosmic travellers experiencing other realms of existence. Love is the key.

*

Following is a comment on purification according to D.K.

“As the soul becomes increasingly potent in the mental and emotional life of the aspirant, it pours in with greater power to the head centre. As we work with our personality, purifying it and bending it to the service of the spiritual will, we automatically raise the energies of the centres up to the centre between the eyebrows.

“Eventually, the influence of each of these two head centres increase, becoming wider and wider, until they make a contact with each other's vibratory or magnetic field, and instantly the Light flashes out. Father-spirit and Mother-matter unite and are at one and the Christ is born. *Except a man be born again, he cannot see the kingdom of God*, said the Christ. This is the second birth, and from that moment vision comes with increasing power".

D.K. also says

In times to come, when humans are more awake and aware, greater peace will reign upon earth. You will realise it is created from within as a result of collective thought and behaviour. The process has begun, and it will be like a tidal wave, clearing psychological debris in its path.

Be prepared for this purifying process to happen in your personal lives. The wave has begun and will not stop. Understanding that the end result will be beneficial will help you to ride the wave with confidence. In astrological terms, transiting Pluto moving into the last few degrees of Capricorn, along with transiting Uranus in Taurus, (two earth signs) will bring the transformational changes needed.

As mentioned previously, the Great Plan for human evolution into the light of higher consciousness is in acceleration mode. The choices you make individually, whether they are unconscious and re-active, or conscious and well considered, will constitute the results experienced.

When aware of the consequences of each choice made, knowing it will affect your life, you make them wisely.

We leave you now to ponder upon our words.

*

The ultimate experience of life on Earth culminates when the Soul, Spirit and personality unite in the body as one. This grand happening occurs when it is realised that there is life **beyond** the psychological process, and that this way of life can be accessed when we have purified our psychology to the extent we no longer react adversely to others' or our own psychological input. We become clear channels of service to the Divine Plan for human consciousness evolution.

Our bodies are a system of intricate biological circuitry designed to know ourselves as an expression of Universal Being. We are a cell in the terrestrial body of the One whose consciousness is now awakening in the human family.

Astrological Ages ago, we humans succumbed to the spell of matter. We fell into this shadow state of consciousness, experiencing amnesia. Like a heavy blanket covering our consciousness, our amnesia kept us separate from the knowledge that we are an aspect of the God Force.

We are now awakening to the fact that we are One with Prime Creator and cannot ever be separated. It is only our minds that believe we have been. The Age of Aquarius is the Astrological Age when this heavy blanket of amnesia is to lift.

We have the choice to accept this fact and open ourselves to greater love and light, or not.

*

As a human race, we are beginning to raise our lower energies into the higher centres. This is a large part of our spiritual evolutionary process. We are in the process of birthing a new galactic human, filled with Christ consciousness. When this is accomplished, we as a collective, can reclaim our status as a valuable member of the galaxy.

*

Conscious Creation

The Arcturians say:

The unveiling of the new system of astrology is a journey of consciousness. The current system was created to maintain a particular level of consciousness and to keep humanity in a state of control and manipulation. This worked for many thousands of years.

Now, times are changing. Creation is speeding up as more and more people awaken to their truth. A wave of self-awareness is sweeping the planet and this wave will continue. It will create many more waves. Life will never be the same. Conscious creation is accelerating. Think something, and that something is created.

New planets are discovered, or rather revealed, when the mass consciousness is ready for their next evolutionary step forward. Then, a bigger picture of creation emerges. The governing bodies of a higher and more refined system reveal themselves. It is this Arcturus Astrology system that is the New Astrology for the New Age. Sirius is the central star.

Within humanity's mind, the story of Isis and Osiris arises to the surface of memory. It is not only the story, but also their teachings that are remembered.

Technology in those ancient times was far more advanced than today. The Source of energy was well known and utilised, mostly for good. However, there was one who chose to use his knowledge for purposes of control and manipulation, thus demonstrating man's power over cosmic forces.

The Universal Law of Karma, of Cause and Effect, never sleeps. Whatever is caused by a misuse of energy, has an effect. Sometimes, the effect is experienced for countless numbers of lifetimes, depending on the type and level of the cause, and the level of consciousness of the perpetrator at the time.

Generally, people are now willing to accept responsibility for their creations. More people are becoming empowered through the truth of their feelings. Yet, many still perpetuate darkness through the strength of their thoughts. Enough people are ready to accelerate their path to the living light of love, and that path is directed by the stellar system with Sirius at the hub of the wheel.

What is above, is also within. There is no separation. All are connected within the great body of The One. What is revealed and perceived without, is the out-picturing of what can be revealed within.

Humans are made in the image and energy patterns of God. We each have the capacity to embrace our God essence and be our God presence. More and more people are choosing this path.

Human nature is naturally curious and seeks to identify and understand the world. Many people will seek to understand the cosmic forces that continually illuminate human experience.

Each of the nine energy centres that orbit around Sirius rule a particular part of the human body. The trinary central star, Sirius, rules the God Presence – the spark that is the seed containing the knowledge and wisdom of All That Is.

When this God seed is nurtured with love, respect and awareness, it sprouts into a healthy plant. Knowledge and wisdom are the branches. One knows, without knowing how one knows. With further nurturing, honouring, great love and consideration, the plant becomes stronger. Its growth radiates to different parts of the body and fuels that part of the body with its essence. The greater the love and nurturing, the healthier is that part of the body. It has always been so.

In ancient times, people lived for hundreds of years. A gradual shortening of life took place in human history in direct relationship to the degree of self-love, self-nurturing and honouring of the internal God presence. When humanity began to worship idols and things outside themselves, human life shortened.

Now that humanity is again beginning to honour, respect, love and nurture their inner God presence, their life span will increase. They will gain more life-force energy and subsequently live healthier, happier and longer lives.

And, so it is and always has been so.

*

God Realisation

The Record Keepers share the two following messages:

As in times gone by, and in times to come, many more people will hear the call of Spirit. This inner call may be faint at first, felt as a longing, or a yearning, for a something. Even though lives will be lived fully, they feel empty. Something is missing.

How does one unite with Spirit?

The process differs with each human being.

Essentially, providing quiet time each day just for self is the first step. Letting go of thoughts during this quiet time is the second. Being in Nature provides solace for the soul and it is easier in Nature to calm the mind.

Committing to doing whatever is necessary to claim your daily quiet "being" time is necessary. This simple daily practice will fuel your soul and provide the quiet space needed to receive your Spirit's guidance. Daily prayers of heart-felt gratitude for all you have, and have learned about yourself, will contribute to the process, as will daily meditation.

An unfulfilling life creates illness. Only you can change your life to one of happiness and joy. These states of being are your birth right. It takes serious intent and daily practice to attain a state of grace.

We leave you now to ponder upon our words.

*

As human thoughts turn to the realisation and acceptance of God resident within their mind and body, life will change. This is the prophesised shift in consciousness that has been long awaited. The next nine years of this decade will be experienced as the Great Awakening, or the Second Coming of Christ. A propitious event that will change human consciousness forever.

All walks of life will experience this shift. As this wave of higher consciousness sweeps through human hearts and minds, major global change will occur. It can be peaceful, or disruptive. A choice only you can make individually and collectively.

God realisation has been proved by scientists as the substance that glues life together, coalescing from tiny particles into frequency waves that permeate all of creation. This essence, as a tangible substance, is in all life, animate and inanimate. As Spirit, it infuses all.

Blessed are we to become consciously aware of this natural state of being. No longer is it only a religious concept, but is a spiritual and scientific fact.

The wave of God consciousness, of spiritual Light, cannot be stopped. It will only increase. The dark will like it not – to no avail.

The awareness of God, or Christ consciousness as a cohesive gel within the hearts and minds of all evolving souls, is life enhancing. It cannot be otherwise. As integration of this knowledge becomes wide spread, human hearts will rejoice.

Soon after will come the realisation that all life is as a cell in the great Body of the One Creator, the Source of the Living Spirit that animates all matter.

Rejoice in this knowing.

We leave you now to ponder upon our words.

*

The Next Human Evolutionary Step

The Arcturians have the following to say:

To some, our messages will have little meaning, falling on deaf ears. To others, they will touch their souls and truth will be felt. It is as It Is.

As the blanket of amnesia is thrown aside more people will awaken and feel the truth within. We persevere.

All humans are at different levels of consciousness. To find others at a level similar to your own can feel as if you are reuniting with a family member, no matter the age difference. You are attracted by each other's inner light.

There is so much love, joy and wonderment yet to be discovered by those pushing aside their amnesia blankets. When the heart is returned to its rightful place in union with Prime Creator, the world shines with radiance and beauty is experienced daily. Joy abounds.

This state of being love awaits all humans willing to move through and beyond dense psychology. This love is found within and is different to the love felt for another human. Divine love rises to the heart centre as the understanding arises from within of the Eternal Presence. This spiritual experience of ignition is felt as a wash of warmth throughout the body and in every fibre of being and can be seen through inner eyes as tiny sparks lighting up.

This is the next human evolutionary step. Those awakening early are the forerunners of a movement that can only accelerate. Part of the Divine Plan for human evolution, the mass awakening cannot be stopped. Many will try and do so. Where there is light there is also dark. The materialising influence can be seductive and addictive. Human minds can become so distracted by it stress and disharmony abounds.

When decisions are made to follow the path of love and joy, stress and disharmony vanish. It takes trust to follow this path. Trust comes through experience.

Love and joy will never let you down.

We leave you now, and will come again.

*

They continue …

As it was in the past, so it will be in your future. Energy is constantly in motion, as with the planets in your Solar System there is continual motion. Energy in motion. Emotion is energy in motion. Thoughts are energy in motion. Planet earth is continually in motion. It may not feel like it.

Energy travels in cycles and spirals. Nature is proof of this motion. Reflect on a sea shell or a snail. Within a human life span are time cycles. Life changes accordingly. Astrology depicts these cycles. The cycles are orderly. You can learn to understand Nature's order. It is constant and forever in motion.

At this point in your evolution, a revolution is taking place. No longer are humans willing to live lives under man-made control systems, created to withhold freedom. In all walks of life, giant corporations are controlling your lives. You have allowed it. It Is.

As many are now waking up to this truth, a quiet revolution is taking place. Old systems of religion, education, politics, finance and health are breaking down. This will continue. As each individual awakens from the deep sleep of amnesia, the ensuing light of awareness will spread. Shadows cannot exist when light penetrates them.

An evolutionary revolution is in effect, and is gathering momentum. Those resistant to change may struggle. Breakthroughs in science will continue, rendering outdated religious beliefs null and void. A new way of being human is emerging. Some will resist this evolutionary spiral, to no avail.

Light will continue to shine in the darkness and the darkness will like it not.

It Is as It Is.

They continue:

All humans were created for the purpose to grow and evolve into super, or Christ conscious beings, aware of their connection to the One Creative Force you know of as God/Goddess/All That Is.

Creation is infinite. As your scientists have discovered, the Universe and everything within it is in a constant state of expansion and growth. Humans are currently in a growth phase. It is accelerating. The growth is related to consciousness.

It is time for humans to become fully conscious beings. Aware of the God Force within and seeking to serve this Force by sharing your unique creations is the pathway. You have free will to express your creations in a way that resonates with your heart and soul.

Each of you has the opportunity to live fulfilling and life-enhancing lives, should you choose to access and perfect your natural creative gifts and talents. Each unique creation contributes to the whole, benefitting all. The design is simple, is it not?

To uncover innate natural creative expression seems to be a problem for some. The uncovering takes self-awareness developed through self-knowledge. To know yourself as a cell within the great Body of the One Source is part of the journey to self-awareness.

This knowing may be intellectual rather than embodied. When embodied, a brighter light shines through the eyes and through the entire persona. Others notice the light and the joy of life that arises from the embodiment of the knowledge.

To live life sustained by sharing one's self-determined unique form of soul expression is the ultimate delight. Life force energy, which is pure love and light, flows from the soul's known and expressed connection to Source, enlivening the body.

Freely given, love heals all it touches. It is the Force that connects all of life. Love really does make the world go around.

*

The Arcturians taught me the new astrology system. My task is to share and teach it. They also taught me to experience, and to embody, unity. We are all One. In the beginning was God, Prime Creator, the Eternal Presence. Everything created, galaxies, stars, planets, humans, animals etc. exist within the energy field of Eternal Presence. No matter what we think, we cannot ever be separated from this field.

Prior to each earthly incarnation, we gather with our group soul to choose the life lessons we need to learn during our forthcoming earthly journey, and the natural gifts and talents we can use to do so. We choose our personality, a one-time composite of soul fragments, in order to facilitate our spiritual evolutionary growth into greater light.

After we identify and transmute our discordant shadow psychology, our spiritual lessons, we return to unity with Source. The astrology system currently in use provides clear, and often self-confronting, guidelines for those dedicated to personal spiritual growth through the realisation, transformation and healing of accumulated karma.

Why did we accumulate karma? The Universal Law of Cause and Effect is always operative. When we create shadow thoughts, consciously or unconsciously, we act in unloving ways. Karma accrues. Density results. Our light dissipates. Our love weakens.

*

I've come to understand Arcturus Astrology as a connection to, or an extension of, Esoteric Astrology. Our expanded organs of perception are needed in order to perceive, understand and work with this system.

The Tibetan, Djwhal Khul, is primarily concerned with the development of consciousness within the human family. This has also been my focus for thirty years. Hence, his request of me to write this book. He must have engaged the Arcturians and the Sirians to assist. My mind boggles and I feel extremely humble.

What a cosmic cook-up! D.K. will probably be having a good laugh at my attempts to de-code.

Following are further findings as at the time of writing:

We humans are transitioning from one consciousness time zone to another, one of much finer frequencies than experienced in our third dimensional world. Beyond the limitations of our Solar System, the Arcturus Astrology System exists in this finer consciousness time zone. Acting as a Light Station and based on light technologies, it creates a stepping stone to other galactic consciousness systems.

As already mentioned, the purpose of our Solar System planets is to activate the unfolding of human consciousness. Astrology is a tool to do so. The planets, as living, evolving organisms are also evolving. For aeons we humans have been spiritually guided to *Know thyself to know God*. When this awareness process is integrated by a majority, group consciousness thought and action come into effect. Unity through diversity.

Bear in mind that our solar system orbits around the Pleiades within a circle called the greater Zodiac, an approximate 25,600 - year cycle. Every planet in our Solar System is linked to the Sun, just as I viewed the Arcturus Astrology system with Sirius as the main central Sun hub linking the nine satellites (constellations or stars) connected by spokes orbiting around the central Sun.

Our solar system is a planetary one. The Arcturus Astrology system is a stellar one, a big difference in frequency. An octave higher than that of our Solar System perhaps? Food for reflection.

D.K continues:

As human thoughts turn to the realisation and acceptance of God resident within the mind and body, life will change. This is the prophesised shift in consciousness that has been long awaited. The next nine years of this decade will be experienced as the Great Awakening, or the Second Coming of Christ. A propitious event that will change human consciousness forever.

All walks of life will experience this shift. As this wave of higher consciousness sweeps through human hearts and minds, major global change will occur. It can be peaceful, or disruptive. A choice only you can make individually and collectively.

God realisation has been proved by scientists as the substance that glues life together, coalescing from tiny particles into frequency waves that permeate all of creation. This essence, as a tangible substance, is in all life, animate and inanimate. As Spirit, it infuses all.

Blessed are we to become consciously aware of this natural state of being. No longer is it only a religious concept, but is a spiritual and scientific fact.

The wave of God consciousness, of spiritual Light, cannot be stopped. It will only increase. The dark will like it not, to no avail.

The awareness of God, or Christ consciousness as a cohesive gel within the hearts and minds of all evolving souls, is life enhancing. It cannot be otherwise. As integration of this knowledge becomes wide spread, human hearts will rejoice.

Soon after will come the realisation that all life is as a cell in the great Body of the One Creator, the Source of the Living Spirit that animates all matter.

Rejoice in this knowing.

*

Creating a New Camelot

I offer you an inspiring vision, to create a new Camelot on Earth. I hope to enthuse you to follow the steps provided. You may stumble and fall occasionally, but I urge you to commit to your journey. You will not be sorry, and you will have all the help you need.

Earth is our home. She provides so well for us, with food, shelter, beauty, bounty and water. Often, we take her for granted. Make a pledge to honour and respect her and to offer her daily heart-felt gratitude. Commit to doing what you can to keep her pristine and healthy.

Commit also to self-transformation through self-knowledge. This process is an internal journey into the dark and the light. The valuable threads you will unravel throughout the journey can reveal your celestial heritage. Also revealed will be the understanding of how you create your reality.

Whatever you place your attention upon will become your reality. You choose the reality you live in. You can do this consciously. You will notice that everyone lives in a world of their own making. Worlds are amorphous, only as real as the attention you place upon them.

I ask you to open your minds and your hearts to experience greater realities. Within your own body lie the answers to the greater mysteries you seek. As you progress through your transformational journey to greater Light and Love, you will notice your body chemical and cellular levels change. This is an indication of your embodiment of greater light.

Imagine yourself as an amplifier of light. As you realise more about yourself and transform darkness into light, the dimmer switch turns, and your internal light becomes brighter. This process leads to the spiritualisation of the physical body. Empower yourself to be the best you can be by reclaiming the Light from within.

We are not only our physical body. We are a spiritual being inhabiting a physical body. Your body is priceless. It contains a tremendous wealth of knowledge. Learn to use the wholeness of your body to understand the deeper mysteries. Learn to decode your knowledge. The best tool I know to aid this evolutionary process is psychological/ spiritual astrology.

You have each been tapped on the shoulder to answer the call to awaken to truth. Many are afraid to traverse the unknown, and to stand for something other than the conventional, the so-called normal. The call is ignored.

It is challenging to move away from your comfort zone of emotional security. However, it is necessary in order to create a new Camelot together, a Golden Age of Peace. There is no judgement on how far you travel because it is your free will choice. I'm hoping to inspire you to heed the call of your soul.

*

Sunlight and Roses

In my recent book, *The Magdalen Codes – Reclaiming your Ancient Wisdom,* I write of my ten-year experiences of being taken, during frequent meditations, to view the alignment of three Suns. Our Sun, the Pleiadian Sun Alcyone, and the Great Central Sun, Sirius. The alignment message was clear. Extra light is bombarding our solar system for evolutionary purpose.

The energy emanating from these three great Lights contain a new set of coded instructions for the spiritual evolution of life and of consciousness. When an organism absorbs Light, it begins to resonate with that organism's DNA. The newly encoded Light entrains our human DNA. This Light can reach us in scaler wave form, as fast as the speed of human thought. It also uploads into the Galaxies.

The times we live in are unprecedented in human history. A great consciousness plan governs human spiritual evolution. We humans, on this tiny outpost in the Milky Way galaxy, are recipients of this plan. We can co-operate with it, or not.

Humanity and the Earth are undergoing a quickening. Science tells us we are moving through a highly energised area of space. All bands within the electro-magnetic spectrum have increased. The Schumann Resonance has had unprecedented spikes. Incoming Cosmic Gamma and

X-rays from far distant galaxies are bombarding Earth, and increasing exponentially.

The Sun is experiencing coronial mass ejections and solar flares that have an impact on humanity socially, economically, physically and psychologically. These incoming energies can create severe weather and an increase in volcanic and earthquake activity. There is also the influx of higher consciousness. Energy that comes from the presence of higher dimensional beings adding to the consciousness energy of the planet.

Some energy healers observe a transformation of the energetic pattern of DNA in their clients although they may not understand what they see. A crystalline pattern of diamond shapes is sometimes seen, as was the case when a medical intuitive observed diamond shapes in my blood. Some see hexagonal shapes, an arrangement of six DNA strands into a hexagon.

These shapes reveal levels of consciousness. There are many mysteries still to solve and they will be made clear when we have the consciousness to understand.

What follows is another mystery I've been given to solve. I wrote about it in Part Two of *The Magdalen Codes.* It involves roses and all they symbolise.

The rose is a symbol of love. It is also the symbol of Mary Magdalen, the feminine Christ. Some medical intuitives have seen a tree of roses in my etheric body.

When I was being trained by the Arcturians to understand the new cosmological system for human evolutionary development, I was given the image of a tree of nine roses around a central larger rose. It was a symbolic model from which I was asked to extract finer details.

The roses morphed into the ten hearts, a symbolic representation of the Arcturus Astrology System. To me, this means that the Feminine Christ is arising within human hearts and minds to take her rightful place beside the Masculine Christ.

The Sirian Goddess is rising.

The Arcturians say:

To understand and work with the new astrology requires a heart full of love and joy. Where there is shadow, our light cannot penetrate. A sacred heart, rejoicing in life and acting on feelings of love, is a state of being. As a small child perceives its world through innocent perception, so too can an adult. Living life in a hectic and polluted environment is not conducive to the attainment of this level of light.

When two people connect through their hearts with no other agenda, a state of wonderment occurs –at the seeming magic of the connection. Their worlds seem to shine more brightly as love flows from one to the other. Their energy fields sparkle, and this light radiance can be seen by those who have eyes to do so.

The love energy created has potency and the potential for good. Love spreads. Others within the vicinity of the loving couple feel the energy and rejoice. It has always been.

Mundanity and structure can take over, and the joy lessens. It need not be so. Self-awareness is required. Isn't love worth sustaining?

Love is the fuel needed to power the human vehicle into top performance. Love requires maintenance and constant fuelling. Love fuel is always available. An open heart, free from structure, is needed to ensure the love fuel functions effectively.

To live life in love requires intention and practice. The mind needs to be mastered because it can destroy the love feelings.

And this is the next evolutionary step for humanity. Evolution into love.

Human beings were created through love to be love. Being love requires the mind to become a servant to the heart.

*

CONCLUSION

As I was about to send this manuscript to the printer, I received the clear message there was something more to add. I was surprised by what follows: An unedited journal note follows:

At sunrise on 6.45 am, 26th January 2009, under a New Moon and Solar Eclipse, 7 degrees Aquarius, I gave birth outside in nature, South East Queensland, Australia, to an etheric Sirian star baby. I was told she was an Avatar from Sirius.

She was light conceived at sunrise on the March Equinox 2008 in the Andes Mountains of Peru at Machu Picchu by the young masculine mountain energy of nearby Wayna Picchu. Following the conception, I was asked to walk around the base of Wayna Picchu to an ancient Moon temple. A challenging walk, where chains and ropes had to be used.

Inside this cave-birthing temple, she received cosmic vibrations from the seven large and extremely heavy orange coloured meteorites, placed in specially carved alcoves. This energy from outer space enhanced her DNA and energy field.

I was guided to sit on a specially carved rock where women in ancient times gave birth to their babies. My footprints added to those already naturally formed by thousands of women birthing their babies in this same spot.

At her Australian 'birth' I was given the information that The Sirian Council of Nine are her Godparents. Each of them gave her a blessing:

From Tara she received the Gift of Divine Wisdom.

Sai Baba blessed her with Cosmic Wisdom and the ability to comprehend the Cosmic Joke.

Yeshua ben Joseph blessed her with the energy of the Divine Masculine,

Beinso Duono with the gift of joy through dance and movement,

Athenia with the gift of beloved sisterhood,

Djwhal Khul with the milk of loving kindness, and manna from heaven,

RA, the gift of loving radiance, warmth and essence of God,

Omega with the gift of loving leadership and authority.

The baby's name is Saiwa, meaning - in an ancient Andean language - the breath of God. My clear guidance was to wear her strapped to my heart and to take her with me each time I visited Sirius and Arcturus. When I next journeyed to Sirius, Lord Maitreya engulfed me with his enormous Light and I felt blessed beyond measure.

*

Now we'll return to the present:

During a recent meditation I asked the Arcturians for a conclusion to my story. Again, The following surprised me.

The human chakra system of spinning energy vortices is a stepping down from Mother Earth's chakras.

The Earth chakras are a down stepping system from the Solar System chakra system, which in turn is a stepping down from a galactic system.

The Arcturus Astrology system is a system of consciousness comprising spiritualised energy from Source, replicated and transduced into all areas of space, finally landing in a human.

When a human becomes conscious and self-aware, the masculine Ida serpent and the feminine Pingala serpent awaken in the human body. As they feel the embodied Eternal Love, the serpents travel up the Sushumna canal, swirling, twirling and meeting at the crossing points of the heart and third eye chakras.

Joyous ecstasy at the reunion with Source is experienced by the human involved. Bursting through the crown chakra the serpents once again embrace their long-separated multidimensional aspects. The Divine Feminine and Masculine unite as one.

The Australian Aborigines, the only remaining human Root race from Lemurian times, celebrate the human attainment of higher consciousness through dance and storytelling. Their most revered feminine goddess is the Rainbow Serpent.

As the etheric masculine and feminine serpents rise from the base chakra they embrace the rainbow colours of the light spectrum, becoming immersed in them. Alive, vibrant, healthy and joyous. Celebration and reunion with the internal God/Goddess the result.

The rainbow children of Light have, in previous lives, done the inner psychological work necessary to prepare for this next stage of human evolutionary development. They enter into this 3D plane to assist their fellow humans do the same. Their internal Rainbow Serpents dance freely and joyously within them. Reflect on the Hopi prophesy in Part Two.

I now share my experience of the Rainbow Serpent. The following is an excerpt from my book, *The Magdalen Codes.*

"Visualise a group of three hundred dedicated spiritual seekers sitting in silence, deep in meditation. I was one of them. During this meditation, at an inaugural Australian spiritual conference in Sydney in the year 2000, we were guided to travel to the great red rock mountain, Uluru, in the centre of Australia. Uluru is a unique global sacred site. This was where the next conference was to take place.

"During this meditation, my inner teacher asked I enter inside the Uluru mountain to be as close as possible to Mother Earth's core crystal. Here I encountered an old aboriginal man who handed me a book of Gaia's records. He telepathically asked I keep them safe until such time as I was guided to share the contents. He said I would know when that time was right. It is now.

"When nearing the completion of the three-day Uluru conference the following year, and while taking a break at a coffee shop, an aboriginal woman approached me. Tina, a complete stranger, was the local 'wise woman'. Her message involved the sacred circle dance Paneurythmy.

"The reason I travelled to Uluru was to teach, at the conference, this circle dance to approximately 250 people. In return for her message, I gave her a copy of my book, *Tara, Emissary of Light*. Glancing at the cover she registered shock. Recovering, she told me she had been shown the same cover image in a vision, two days previously.

"During the afternoon of the last day of the conference I was given a second, even more powerful message. Tina had directed the tribe's senior representative, Uncle Bobby Randall, to take me to the Heartland, some kilometres from Uluru. Uluru, in the centre of Australia, is a huge red rock monolith and considered Australia's most important sacred site. I agreed to accompany him the following day.

"Collecting me early in the morning, he drove along the main road for some time until suddenly veering off the road into bushland. Driving along a bumpy bush track he slowed to show me the shed where his mother had lived and been born. Instinctively and spontaneously, I asked him to stop the truck, feeling the need to perform a thanksgiving ceremony at the shed.

"Following the ceremony, Uncle Bobby drove on slowly over the unmade track giving me time to connect to the land. Out of the corner of my eye I noticed, some distance ahead, a high rocky ridge. Driving on further for maybe five minutes Uncle Bobby suddenly stopped. He said I was to climb this same rocky ridge to attend to spiritual women's business.

"I had no idea what he meant. He suggested his wife, Hazel, accompany me, saying he would come back when the time was right. We watched him drive away into the distance. We didn't see any other sign of habitation or human life.

"Hazel and I began to climb. Close to the top of the ridge I felt intense pulsating energy activating my heart, so I began breathing deeply into it. The energy gathered force, as did my breath. I began to feel emotional as unconscious memories awakened, flooding my body. When the powerful energy reached my throat deep penetrating sounds emerged that became heartfelt sobs. Soul memories consumed me. My conscious mind could not comprehend.

"I was telepathically guided to lie down on the narrow strip of red soil at the top of the ridge, with my face to the ground. My entire body began to heave and writhe. The primal sounds became deeper and more intense as the waves of overpowering internal energy gathered force. Stronger and more powerful sounds erupted through my throat.

"Inner guidance directed me to place the red soil over my hands, face, and any other exposed parts of my body. The instant I rubbed the red soil on my face I felt a penetrating heart-felt and ancient soul connection to this land, and to Gaia, Mother Earth.

"More primal sounds erupted as I connected with her. Feeling myself as the rainbow serpent, I began to writhe, twist and slither along the ground. I moved, face and body down, hands grasping the soil, sobbing deeper with even more profound sounds. Time ceased to exist. Then, I felt myself giving birth.

"As the rainbow serpent, I experienced birthing many human souls into Mother Earth's warm embrace. I knew then that the First Australians had not walked across the Bering straits, nor did they come to Australia from other countries. They were birthed here, into this land. Time seemed to stand still. Energetically and emotionally spent I was unable to move, still immersed in the reality of the powerful experience.

"Eventually I sensed Hazel sitting behind me, patiently and silently supportive. Much later she rested her hands on my shoulders and helped me to sit up, asking if I could manage to scramble down the steep slope with her help. She could see Bobby's truck returning.

"Unsteady on my feet but with Hazel supporting me, we climbed slowly and cautiously down the embankment. Spent, stunned and completely depleted of energy, I was unable to speak. It was the most powerful experience I have ever had. Uncle Bobby picked us up and drove back to Uluru. When he dropped me at my hotel I sat, still stunned, on a chair in the courtyard garden, covered with red dirt, where I remained for hours.

"Sometime after midday, Hazel returned asking if I wanted to talk about the experience. I shared, saying that one of the deepest emotions I felt was loss. The loss suffered by humanity of the wisdom and knowledge of the primal archetypal force behind creation, the Mother Goddess.

"Hazel told me the Rainbow Serpent goddess, considered sacred, is so steeped in mystery, secrecy and magic she is not spoken about. She is the ultimate Creator, the feminine aspect of God. I felt blessed, and also humbled to have had the experience of being her, birthing humanity. I went to bed very early that night.

*

The nine starry satellites written about above are the galactic chakras that replicate energetically on planet Earth, and in the energy field around the human body. Each of the nine spin around a vortex, the group unity of the Council of Nine from Sirius, uniting through their loving service and commitment to Prime Creator's idea. It is one unit of group consciousness vibrating in tune like an orchestra, all senses activated. All aligned in the one purpose. To do what they can to raise human consciousness.

*

My experience at one of planet Earth's chakras, Uluru in Australia, was of the etheric Rainbow Serpent giving birth to a more evolved human race. According to D.K., - the sixth Root Race. How long the physical manifestation takes is uncertain. It depends upon the human collective mind.

Words from D.K:

Remember when I asked you to write a book for us? I am the spokesperson for a group of celestial guides, and you agreed saying you could not do intellectually, as Alice A. Bailey had done.

Hmmm, I thought to myself, what's coming?

You must admit we have acted according to your request. It's been quite a journey. Three books later, and many more to come. We have enjoyed the experience, as you have. That is all for now.

*

Oh Ashtara,

We see and feel your joy as you crack the final code of your book writing. A joyful future lies ahead. A transition from a teacher to a story teller has now taken place. We congratulate you an enormous undertaking and a successful conclusion. Your book will open the hearts and minds of many readers.

Dance lightly through life and it will be joyous.

We are the Arcturians and we will come again.

*

ACKNOWLEDGEMENTS

With a heart full of love and gratitude I thank The Tibetan, Djwhal Khul, (D.K.,) for asking me to write a book for him. What a growth journey it's been! And a huge "thank you" to the Council of Nine from Sirius and the Arcturians for their total trust and faith in me. Sometimes when writing this book, I would feel emotionally overwhelmed when I felt into their trust, and was unable to function for hours. Like a stunned fish, I waited until I come to my senses. I didn't allow the responsibility of my service work to get me down. I just got on with the job and did the best I could.

Sometime in the early 1990s, the Arcturians, my friends and galactic playmates, entered my meditative reality, asking me to be their scribe. I agreed. Their first transmission was a small book about human evolution on earth titled *Gaia, Our Precious Planet*. I transcribed it word for word, and it became a popular paper book.

But who are the Arcturians? I hear and feel them to be loving, patient and caring higher-dimensional light beings. They are clear and direct with their telepathic communication and their transmissions are often delivered at the most unexpected times, usually in the early hours of the morning. They transmit valuable information I record and share.

From my understanding and experience, the Arcturians exist in a higher dimensional frequency to that of humans. Our human perception of reality is completely different from theirs. Their reality is based on divine love and the light of higher consciousness. They say there is always a bigger picture behind our reality, and it's for our highest good we learn to focus on it. They trained me to experience and to understand the cosmic system guiding human consciousness evolution I share with you.

My heart fills with gratitude for the Council of Nine from Sirius who trained me to develop my extra-sensory perception and to open my heart to divine love. For many years they guided, trained and healed me whenever it was needed.

And thank you to my celestial guides, without whom my journey to the light may not have taken place.

On this third dimensional plane, I acknowledge a group of students and friends who, with love in their hearts and the desire to serve, gathered at my home on Sunday morning the 19th November 2018. Our intention, and our purpose was to anchor into Gaia's etheric field the Arcturus Astrology trinary system for human consciousness development.

I sincerely thank this special group of people who believed in me, and in the Arcturians, and who assisted me in maintaining and sustaining the energy needed to do our work. It was extraordinarily powerful morning.

I'd also like to thank another group, all women who, four months later, under a Full Moon in Libra on Good Friday 2019, ceremonially and physically anchored into the Earth at Woolumbin, an ancient aboriginal sacred site in northern New South Wales, Australia, the energy of the Arcturus Astrology System. They came to my Easter Retreat from different Australian states, and I met many of them for the first time. They also believed in me, and in the Arcturians. Thank you so much Retreat playmates. When integrated into the physical third dimensional plane this cosmic system enables the birthing of a new Camelot, a Golden Age of peace.

And, to my students who, unaware of the impact of Arcturus Astrology on themselves and their lives, chose to attend my first Arcturus Astrology course, conducted one day a month for ten months. You are brave and curious souls indeed.

Thank you to Pauline Alcock, Jude McDougall, Anna Neumann, Wendy Davies, Jody Campbell, Mary Hardwick, Ziva Zavadil, Kim Morotti, Barbara Davies, Shayne Hertzberg, Joanne Bacon and Louise Fewtrell for believing in me and in my work. May your precious hearts remain open and your light continue to shine upon those in your circles.

Finally, and with deep gratitude I acknowledge a kind and generous neighbour and friend, Henry Koster, who spontaneously offered to edit this book one early morning when I was walking my dog. I accepted. Carefully, and with impressive Virgo detail, he did so. He also offered valuable suggestions for which I'm exceedingly grateful. Thank you so much Henry. You would have had no idea of what you were letting yourself in for. I hope you enjoyed the ride.

And to my readers. Thank you for purchasing my books. My hope is the content will open your hearts and minds to a reality beyond this third dimensional plane, a reality your soul knows and yearns to re-experience.

Many blessings and much love,
Ashtara

Milton Keynes UK
Ingram Content Group UK Ltd.
UKHW041502111223
434169UK00001B/22